Design and Manufacture of Plastic Components for Multifunctionality

PLASTICS DESIGN LIBRARY (PDL)
PDL HANDBOOK SERIES
Series Editor: Sina Ebnesajjad, PhD (sina@FluoroConsultants.com)
President, FluoroConsultants Group, LLC
Chadds Ford, PA, USA
www.FluoroConsultants.com

The **PDL Handbook Series** is aimed at a wide range of engineers and other professionals working in the plastics industry, and related sectors using plastics and adhesives.

PDL is a series of data books, reference works and practical guides covering plastics engineering, applications, processing, and manufacturing, and applied aspects of polymer science, elastomers and adhesives.

Recent titles in the series

Biopolymers: Processing and Products, Michael Niaounakis (ISBN: 9780323266987)
Biopolymers: Reuse, Recycling, and Disposal, Michael Niaounakis (ISBN: 9781455731459)
Carbon Nanotube Reinforced Composites, Marcio Loos (ISBN: 9781455731954)
Extrusion, 2e, John Wagner & Eldridge Mount (ISBN: 9781437734812)
Fluoroplastics, Volume 1, 2e, Sina Ebnesajjad (ISBN: 9781455731992)
Handbook of Biopolymers and Biodegradable Plastics, Sina Ebnesajjad (ISBN: 9781455728343)
Handbook of Molded Part Shrinkage and Warpage, Jerry Fischer (ISBN: 9781455725977)
Handbook of Polymer Applications in Medicine and Medical Devices, Kayvon Modjarrad & Sina Ebnesajjad (ISBN: 9780323228053)
Handbook of Thermoplastic Elastomers, Jiri G Drobny (ISBN: 9780323221368)
Handbook of Thermoset Plastics, 2e, Hanna Dodiuk & Sidney Goodman (ISBN: 9781455731077)
High Performance Polymers, 2e, Johannes Karl Fink (ISBN: 9780323312226)
Introduction to Fluoropolymers, Sina Ebnesajjad (ISBN: 9781455774425)
Ionizing Radiation and Polymers, Jiri G Drobny (ISBN: 9781455778812)
Manufacturing Flexible Packaging, Thomas Dunn (ISBN: 9780323264365)
Plastic Films in Food Packaging, Sina Ebnesajjad (ISBN: 9781455731121)
Plastics in Medical Devices, 2e, Vinny Sastri (ISBN: 9781455732012)
Polylactic Acid, Rahmat et al. (ISBN: 9781437744590)
Polyvinyl Fluoride, Sina Ebnesajjad (ISBN: 9781455778850)
Reactive Polymers, 2e, Johannes Karl Fink (ISBN: 9781455731497)
The Effect of Creep and Other Time Related Factors on Plastics and Elastomers, 3e, Laurence McKeen (ISBN: 9780323353137)
The Effect of Long Term Thermal Exposure on Plastics and Elastomers, Laurence McKeen (ISBN: 9780323221085)
The Effect of Sterilization on Plastics and Elastomers, 3e, Laurence McKeen (ISBN: 9781455725984)
The Effect of Temperature and Other Factors on Plastics and Elastomers, 3e, Laurence McKeen (ISBN: 9780323310161)
The Effect of UV Light and Weather on Plastics and Elastomers, 3e, Laurence McKeen (ISBN: 9781455728510)
Thermoforming of Single and Multilayer Laminates, Ali Ashter (ISBN: 9781455731725)
Thermoplastics and Thermoplastic Composites, 2e, Michel Biron (ISBN: 9781455778980)
Thermosets and Composites, 2e, Michel Biron (ISBN: 9781455731244)

To submit a new book proposal for the series, or place an order, please contact David Jackson, Acquisitions Editor
david.jackson@elsevier.com

Design and Manufacture of Plastic Components for Multifunctionality

Structural Composites, Injection Molding, and 3D Printing

Vannessa Goodship
Bethany Middleton
Ruth Cherrington

ELSEVIER

Amsterdam • Boston • Heidelberg • London
New York • Oxford • Paris • San Diego
San Francisco • Singapore • Sydney • Tokyo
William Andrew is an Imprint of Elsevier

William Andrew is an imprint of Elsevier
The Boulevard, Langford Lane, Kidlington, Oxford, OX5 1GB, UK
225 Wyman Street, Waltham, MA 02451, USA

British Library Cataloguing-in-Publication Data
A catalogue record for this book is available from the British Library

Library of Congress Cataloging-in-Publication Data
A catalog record for this book is available from the Library of Congress

ISBN: 978-0-323-34061-8

For information on all William Andrew publications
visit our website at http://store.elsevier.com/

Publisher: Matthew Deans
Acquisition Editor: David Jackson
Editorial Project Manager: Peter Gane
Production Project Manager: Nicky Carter
Designer: Mark Rogers

Typeset by Thomson Digital
Printed and bound in USA

In memory of my fantastic dad.

He taught me that there is no such thing as too much caffeine or too much cake.

The authors would also like to dedicate the book to the following people: Rebecca Williams, Tony Middleton, Adam Longfield, and Deb Bate.

GO PAWS TEAM!

Contents

List of Contributors

All contributors are located at: Warwick Manufacturing Group,
University of Warwick, Coventry, UK

Ruth Cherrington

Vannessa Goodship

Jianwang Liang

Bethany Middleton

Benjamin M. Wood

Preface

While plastic manufacturing continues to be driven by the market pull, new developments in materials and processes have opened up new design possibilities for manufacturers to add value and function to their products. Mass customization and linked systems of material suppliers, manufacturer, and end user also allow more control within both production and supply chains than ever before.

Multifunctionality not only provides an enabler for manufacturers to create exciting and distinct products in the marketplace but also allows greater control of in-house manufacture. One component can replace what may have been a number of different components that were procured and assembled separately.

Multifunctionality can be brought about in three distinct ways: by embedding in material, by adding extra layers and function (addition), or by integrating multifunction into the entire structure.

The book explores each of these concepts throughout the book, at a level that should be accessible to manufacturing engineers, designers, and students interested in the manufacturing area.

Chapter 1 introduces what is meant by multifunction and presents examples along with introduction to some potential plastic process routes not dealt with elsewhere. Chapter 2 looks at the principles in "embed," by introducing some multifunction materials. It also begins exploration of the concept of "addition" by looking at deposition methods for adding materials to substrates such as inkjet printing. Chapter 3 moves these concepts into bulk manufacturing with a first look at composite materials and manufacturing methods. Chapter 4 focuses on the mass manufacturing method of injection molding and how multifunctionality has slowly developed as technology has advanced. Chapter 5 then looks at the design possibilities in the highly publicized additive manufacturing process. These two processes are chosen to provide two very distinct approaches to manufacture. One process is designed for mass manufacture and one for highly customisable deigns, at the interface across manufacturing is a platform for highly multifunctional design. Finally, Chapter 6 briefly ties some of these concepts together and closes the book.

I hope you enjoy exploring the possibilities.

Dr Vannessa Goodship

Acknowledgments

Many thanks to Colin Tirel and his colleagues at ARBURG, Germany, for providing the pictures in Chapter 4.

1 Introduction to Multifunctionality and Manufacture

Ruth Cherrington and Vannessa Goodship

1.1 Introduction to Multifunctionality and Manufacture

Multifunctional materials are designed to improve overall system performance. The term "multifunctional" is commonly used to describe something that has or fulfils several functions. Multifunctional materials can be classified and grouped into a wide range of categories and the range significantly depends on materials and applications.

There are several ways in which the word "multifunctionality" can be interpreted, these have been categorized into three main areas (Matic, 2003).

1. Embedding of functions within a component – these materials comprise two distinct phases where one function is embedded in another. Multifunctional composites often fall into this category as a variety of components are embedded within a common part.
2. Addition of functions into the system – the introduction of a material to provide additional performance. This can be introduced by a simple coating or lamination process; an example of this is surface-mounted health monitoring.
3. Integration of functions shared in a volume of material – these types of materials are fully integrated. There is little or no physical distinction between them and the phases are intermeshed. These types of systems are often developed to reduce volume or mass. An example is an embedded antenna or battery that shares the mechanical load.

Nanomaterials, biomaterials, and "smart" materials (those which change to and external stimuli such as temperature, pressure, or light) are being incorporated within traditional structural materials to produce performance-tailored products. These include ceramics, polymers, and metals (which will be discussed in more detail in Chapter 2). These speciality materials are being tailor-made to provide specific functions, which are finding homes in a wide range of applications.

Design and Manufacture of Plastic Components for Multifunctionality
http://dx.doi.org/10.1016/B978-0-323-34061-8.00001-6

Although as we have discussed there are several ways in which "multifunctionality" can be interpreted and therefore integrated into materials, one of the simplest methods is by simply embedding of functions within a component (Matic, 2003). Nanomaterials possess unique, beneficial chemical, physical, and mechanical properties that have been successfully incorporated within components for a wide range of applications. One particular area that is likely to continue to expand is the use of nanomaterials as functional fillers in a polymer matrix. Nanoscale materials differ significantly to that of their corresponding bulk counter parts offering several advantages including increased surface area, decreased sintering temperatures, improved photocatalytic activity, and antimicrobial properties.

There are several examples of how nanomaterials have been successfully incorporated into products:

- Plastic beer bottles that contain carbon nanotubes (CNTs) to stay cold for longer periods of time (Arivalagan et al., 2011)
- Tennis racquets reinforced with carbon nanotubes to improve the mechanical properties of the composite (Hayashi et al., 2007)
- Nanoclays platelets are incorporated into food packaging films to provide a barrier against oxygen and water; therefore, extending the shelf life (Goldberg et al., 2011)
- Silver nanoparticles are embedded within plastics to kill bacteria in refrigerator liners (Kampmann et al., 2008)

There are a number of ways nanomaterials are incorporated into a composite, but on a bulk scale this is done by the use of the plastic extrusion processes (which is discussed further on). One of the main problems associated with the use of nanomaterials in composites is the creation of a uniform dispersion of particles within the polymer matrix. Nanoparticles have a tendency to cluster together, due to large surface forces on the surface of the particles. These are known as either aggregates or agglomerates, depending on the nature and strength of the bonds between the particles. These aggregates or agglomerates are an undesirable formation and a uniform dispersion is needed to produce consistent and reliable properties within the composite. Research has found that the addition of only 0.5 wt% CNTs in a polypropylene matrix results in the formation of large aggregates (Tjong et al., 2007); therefore, modifications to conventional processes will need to be realized to fully utilize materials such as these. CNTs are not unique in this respect, with innovative new materials come several challenges associated with the manufacturing, processing, and incorporation of the materials into usable products. Therefore, this book will look at some of the challenges and trends in design, integration, and manufacture in high-volume (injection molding) and low-volume (composite and 3D printing) applications. It will also look at other common process routes in relation to the functionality trends being exhibited in these processes.

As the drive for automation and customization of both low volume and mass production continues, increased multifunctionality in products has much to offer both manufacturers and wider society targets in enabling the future "internet of things," energy generation and energy efficient products, to create a more sustainable society. The popularity of the number of integrated and hybrid process routes continues to grow and manufacturers can now choose to process and integrate not only multimaterials of the same type but also multiple material families (metal/polymer/ceramic/wood), thereby removing finishing processes such as assembly, painting, and in some cases entire traditional manufacturing routes.

In the future, it is therefore likely functions such as energy harvesting (be it solar, kinetic, etc.), energy storage (batteries), sensors, and data capture and transmission devices will become common place, even on the most ordinary of consumer products. There is no doubt that continued breakthroughs in nanotechnology and plastic electronic applications have a major role to play in this future, but it is the manufacturers who design, manufacture, and sell such products that will ultimately enable society to move in this direction. However, polymer processing technology is often overlooked in such developments, and it is therefore exploring this technology that is the focus of this book.

For example, injection molding is the most popular process used to mass produce plastic components. The process has evolved responding to the demand for added value and high performing components, so multimaterial moldings are produced to provide that added value through extra functionality. An illustrative example is that of the multicomponent injection molding process, where two or three polymer materials are injected into the mold to create a layered structure (Goodship and Love, 2002). This process is used for a wide range of applications including the use of injecting a layer of conductive material to provide electromagnetic interference shielding for electronic products. Although injection molding is a versatile process, it is essential that the processing parameters are optimized to prevent undesirable defects (these along with a closer look at the processing options will be discussed in more detail in Chapter 4).

Polymer-based structural composite materials are used for a wide variety of applications from wind turbine blades, bridges, construction, weight-bearing automotive components, and aircraft wings to name just a few application areas. The ability to place function within such long-lasting structural components for health monitoring has long been an active research target; more recently, integrating power sources such as structural batteries or harvesting energy by integrating functions have become of interest as the needs of the energy sector and the potential within the materials sector have become a more focused research target.

Therefore, composite processes that have remained relatively unchanged are becoming increasingly hybridized by the integration and use of other complimentary materials and processes. This will be explored further in Chapters 2 and 3.

3D printing has seen significant interest in both the research community and wider general public awareness over the last few years, which has enabled the development of printers capable of processing a wide range of materials. The main advantage of this process is the capability to produce parts without the need for expensive tooling or excessive material. Another useful advantage is the layer-by-layer deposition technique provides the ability to quickly change between materials or incorporate complex multimaterial designs with the process. This provides a huge potential for the integration of conductive tracks with plastic products to provide a lightweight, flexible, customizable alternative to circuit boards used in most electronic products (Kirleis et al., 2014). Chapter 5 will discuss the applications of 3D printing, advantages, and disadvantages of the technique and some of the research trends with respect to design and manufacture of multifunctional components.

While specific chapters of this book will focus on three main and distinct process routes, for composite, injection molding, and 3D printing, there are of course a large number of other process routes. The reason to focus on these three as standalone chapters and not, for example, blow molding is that the processes for functionality covered do overlap across other processes. Those interested in blow molding will find examples in both Chapters 2 and 4 of applicability to this process. Therefore, a brief coverage of other processes will now be given with details of applicability that can be found within the book in other chapters.

1.2 Plastic Processing Routes

As stated for composite, injection molding, and 3D printing processes, the reader should go to those specific chapters. What now follows is an introduction to multifunctionality for "other" processes beginning with extrusion-based processes. Where links to other processes exist these will be highlighted within the specific sections and reference to Table 1.1 should

Table 1.1 Plastic Processing Processes and Their Applicability to Book Chapters

Process Route	Chapters	Other Applicability
3D printing	5	2, 3, 4
Blow molding (extrusion)	1	2, 4
Blow molding (injection)	1	2, 4
Low volume composites	3	2, 4, 5
Extrusion (single screw)	1	2, 3, 4, 5
Extrusion (twin screw)	1	2, 3, 4, 5
Compression molding	3	2, 4
Injection molding	4	2, 3, 4, 5
Lamination	1	2, 3, 4
Roto-molding	1	2
Sheet and film production	1	2, 3, 4
Thermoforming	1	2, 3, 4

also be used to guide the reader to relevant areas as understandably there is considerable crossover and therefore subjects are generally addressed in one main chapter rather than repeated in all. An example of such a cross over technology is extrusion that is now introduced.

1.2.1 Extrusion-Based Processes

1.2.1.1 Single and Twin Screw Extrusion

The fundamental operation of an extruder, that is to feed, mix, and homogenize a material, forms the basis for many other plastic processes. Extruders are also used to produce the raw materials and compounds for other major processing operations that are the main focus of the book chapters. Extruders come in two types, a single screw extruder that is gravity fed and a twin screw extruder that needs metered feedstock. The difference in the two is both the tolerance to loading with nonpolymeric materials and the level of mixing that is possible within each system. This is ultimately what defines which of the two systems is necessary. Intimate mixing of high loaded materials will generally require a twin screw machine. To add low loadings of materials where dispersion is not critical (e.g., a color, or a percentage of recyclate of the same material), a single screw machine is normally more than adequate. For further details on extrusion technology, the reader is referred to the specialized texts found in the Further Reading section.

A screw system and its modifications can be found not only on extruders but within the blow molding process and the injection molding process. Extruders themselves can also be used to produce a variety of finished and semifinished goods such as pipes, profiles, thick sheets, and thin films by modifying the shape of the exit die. Utilizing coextrusion considerably expands the multifunctional potential of extruders and they can be integrated with other processes, for example, to coat wire or produce netting-covered materials. Given that extrusion provides the feedstock for the other thermoplastic processes discussed, consideration must also be given to Chapter 2 and the potential of extruded formulations to be inherently functionalized through the addition of any number of materials. The limitations on materials come down to two simple properties; to be useful they must be able to be fed though a screw, homogenized with the polymer, and exit the die with the polymer matrix flow (this is a practical size issue, they must be small enough so as not to block the die), and they must be both heat and shear stable to survive this process in a useful form. Beyond that, there is very little to prevent materials being added and therefore a whole range of metal, ceramic, carbon, wood products-based materials have been successfully extruded within a polymeric material. Therefore in producing feedstock materials for other processes, be it a compound for blow molding or a multilayered sheet for subsequent thermoforming, there is little to contain an imaginative designer within this process. A basis schematic of the process is given in Figure 1.1.

1.2.1.2 Sheet and Film Extrusion

What exactly separates the term a "sheet" and a "film" is widely arbitrary; both terms are used interchangeably in the plastic industry. Within this book the word sheet should be considered to relate to a large volume of plastic with a thickness >5 mm and a film a thinner section <5 mm. A number of processes either directly utilize sheet materials (thermoforming) or sheet and film materials can be integrated indirectly (injection molding, blow molding, compression molding, vacuum bagging, etc.).

These films are generally made by

1. Blown film extrusion using an annular die and inflation
2. Cast extrusion using a slot die

A third nonscrew-based method for material combinations not possible by these processes is lamination that will also be covered later in this section.

These types of processes have great relevance for the food packaging industry and particularly in the relatively new field of active packaging to

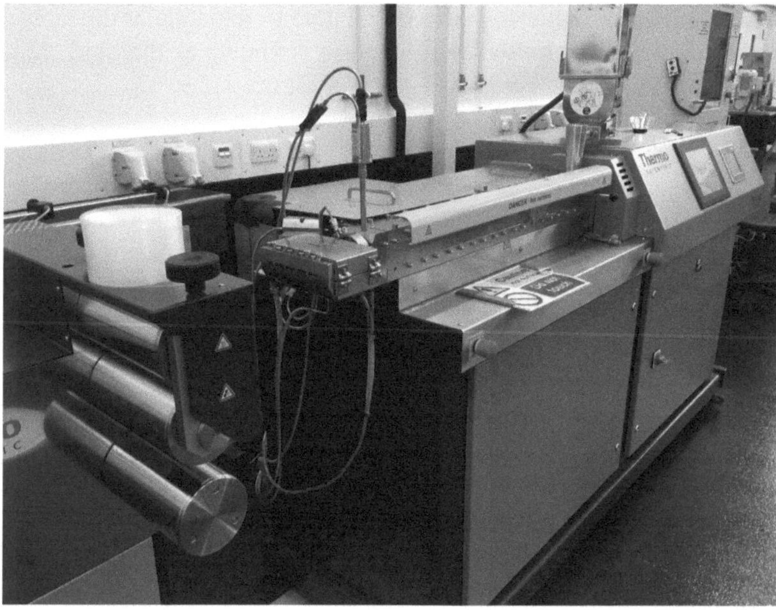

Figure 1.1 Extrusion.

preserve and extend the life of foodstuff. An example was given earlier in the use of nanomaterials in food packaging films to provide a barrier against oxygen and water, therefore extending the shelf life.

A further example area of current research focus was also highlighted – the hygiene of a surface to prevent bacteria forming or to kill it. Silver nanoparticles are embedded within plastics to kill bacteria in refrigerator liners. These kinds of technologies can be applied across a wide range of process routes, but these particular examples apply to film and sheet manufacture. For more detail on similar packaging applications, the reader is referred to the Further Reading section.

1.2.1.3 Film Blowing

At the basis of film blowing is a hopper fed single screw extruder through which polymer feedstock enters the extruder throat to be heated and homogenized to create a molten material. At the exit to the extruder is an annular die to form a tube of material (as in blow molding), but in this case the tube is fed upward into nip rollers that seal the tube allowing inflation by an air ring positioned inside the tube. As the tube inflates, its thickness becomes uniformly thinner as orientation in the transverse direction occurs, allowing a variety of thicknesses to be produced by combined control of the extruder throughput, inflation, and roller speed.

As the inflation diameter increases relative to the annular die size, material orientation is increased and this ratio is known as the blow ratio. A simple schematic of the process is given in Figure 1.2.

1.2.1.4 Cast Extrusion

Cast extrusion can be used to produce single materials (monolayer film) or as a coextrusion to produce multilayer films. Each individual layer requires an extruder, which is combined in a special manifold die that feeds the laminar and separate material streams together. Like blow molding (that follows), it is necessary to use tie layers to stick any incompatible materials together.

1.2.1.5 Blow Molding

Blow molding is widely used to create hollow objects. As well as commonly seen consumer items such as carbonated drink bottles, plastic milk containers, and shampoo bottles, this process is also used for plastic petrol tanks and to create larger technical items such as automotive spoilers.

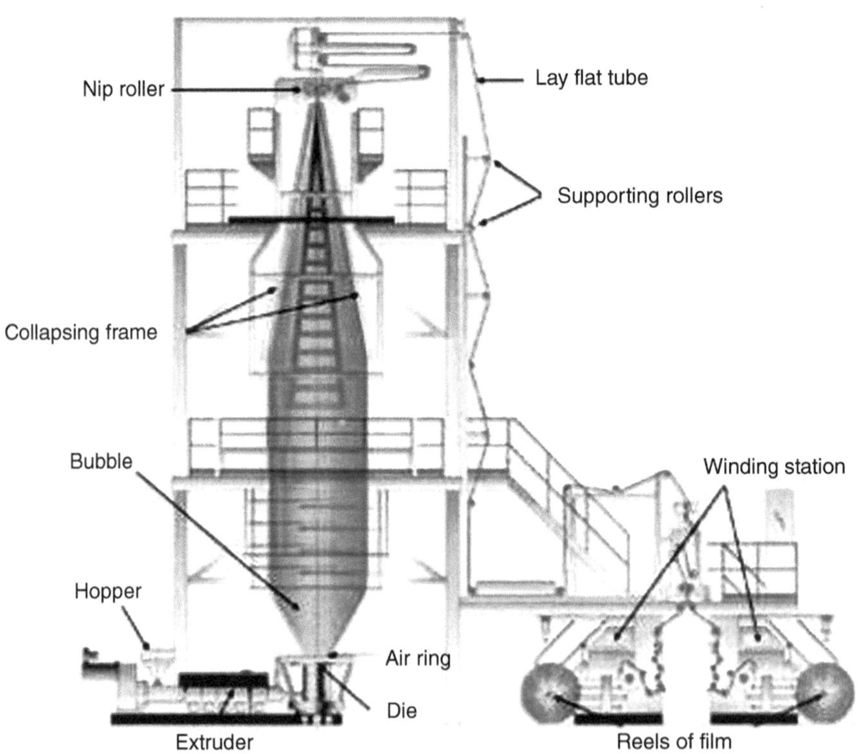

Figure 1.2 Film blowing.

Blow molding can be split into two distinct process routes, namely injection blow molding and extrusion blow molding. The distinction in process is best illustrated on the components themselves. Those made by injection blow molding do not have the bottom seal line that can be seen on the bottom of extrusion blow moldings. (This can be simply seen by comparing the bottom of a carbonated drinks bottle – injection blow molding, no visible seal line, and a plastic milk carton – a visible seal line). This difference in process means the options for multifunctionality are different and therefore the processes will also be treated distinctly here.

1.2.1.5.1 Extrusion Blow Molding

Extrusion blow molding can be operated as either a continuous or semicontinuous process depending on the requirement. In extrusion blow molding a plastic material feedstock is fed through a hopper onto a feeding transfer screw. This feeds material into an especially designed blow molding die that directs the material flow to create an extruded hollow tube called a parison. This parison drops with gravity until it is a process-specific length, which triggers the mold to close around it and seal the bottom edge (that gives the distinctive end seal). A cutting knife slices the open top of the parison and a blow pin moves to the top of the tool and inflates the parison against the walls of the tool cavity. The cooled tool walls solidify the molding and after a preset time it is ejected and the process repeats with the newly extruded parison. A schematic is shown in Figure 1.3.

Blow molding is a relatively low-pressure process in comparison to injection molding, but like injection molding it offers considerable opportunity for the use of inserts. However, unlike injection molding a larger downside is the high scrap level inherently produced by blow molding (which is easily and routinely fed back in as regrind) as a result of parison cut-off. Regrind quality can be compromised if inserts contaminate the process. Therefore, postassembly of inserts may still be preferred in many cases.

Due to the hollow nature of blow moldings, they can also act as skin materials to be filled by other materials such as foam for crash protection. In-mold labeling (IML) is also commonly utilized in blow molding processes and will be discussed in Chapter 4.

1.2.1.5.2 Coextrusion Blow Molding

If more than one material is used to make up the parison, the process is termed coextrusion blow molding – again, there are the analogous coextrusion and coinjection (Chapter 4) processes. The use of coextrusion offers considerable potential for multifunctional design. A commercial example

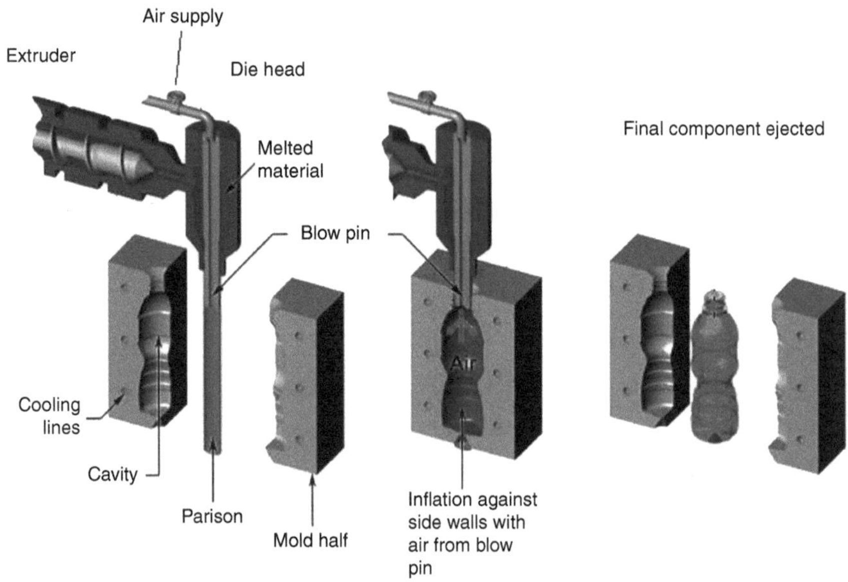

Figure 1.3 Extrusion blow molding.

of this is the production of an automotive fuel tank that can be made of five (or more) layers of material.

An example structure of this would be HDPE/adhesive/nylon/adhesion/HDPE. The HDPE provides the mechanical integrity and water resistance in this structure, while the nylon provides resistance to gasoline vapor permeation. A similar structure also provides a solvent barrier (nylon layer) for pesticide bottles. The adhesive layers are required as HDPE and nylon do not naturally stick to one another and therefore a further material is required to bond them together. These layers are called "tie layers" and need to be heat stable and extrudable for use in both coextrusion and coextrusion blow molding applications. Figure 1.4 illustrates a coextruded bottle material structure.

A number of commercial tie layers are available; an example tie layer for these indicative applications is Dupont Bynel® 4104 (Lee, 2011). Tie layers ideally also need to be recyclable with the bottle production waste stream to allow the regrind to be recycled and fed back into the system for efficient material use and cost control. For this reason, further regrind layers added to the functional structure are also commonly found.

It should be remembered that each separate material needs its own individual hopper unit, and the die manifold, therefore, also becomes more complex to design; however, the design possibilities increase. For

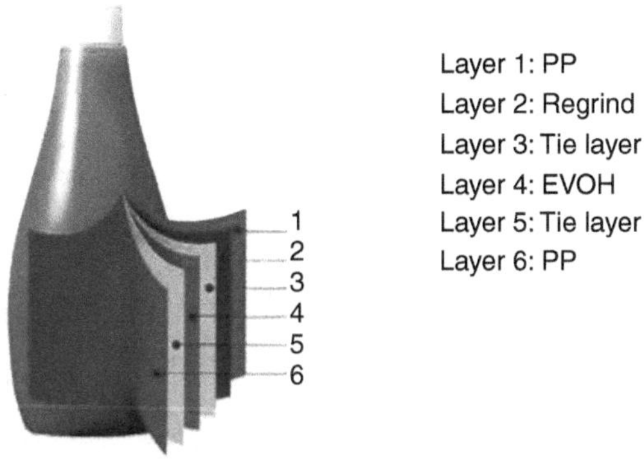

Figure 1.4 Coextrusion material structure.

example, with the two earlier examples of active packaging there is now the potential to add further functionality to the component by the addition of components to each individual material feedstock as long as adhesion and integrity of the layers is considered. Using just the two examples already introduced here, you could theoretically produce a bottle by coextrusion to preserve and extend the life of the contents with an inner barrier layer and then sterilize the fingers of the user of the bottle by incorporating an antibacterial layer on the outside of the bottle.

1.2.1.5.3 Injection Blow Molding

The best known examples of injection blow molded components are clear polyethylene terephthalate (PET) bottles used for carbonated drinks. In contrast to extrusion blow molding, injection blow molding is a cyclic noncontinuous process.

Injection blow molding can be split into three distinct phases that create an injection blow molded component:

1. The injection molding of the preform component onto a support pin/rod
2. The inflation of that component through the support pin
3. The removal of the blown component

While these operations can be and are performed together, the injection molding and blow molding can also operate as two very distinct operations. The advantage of this process over extrusion blow molding is the low scrap levels produced for each container, but with two distinct processes, the tooling costs are much higher as both an injection molding tool and an inflation station are required.

Injection molding itself is dealt with in Chapter 4, so little will be added here on the process other than this is where the potential to multifunctionalize the materials in this process must be done. Performs can be created as multilayers and therefore functional layers can be introduced, as will be further discussed in Chapter 2. Designing for the actual performance of a preform, once it is also inflated, is a highly complex issue in itself and will not be further discussed here.

Once a preform is produced, it then needs to be inflated through the core rod to the shape of the blow mold cavity, cooled and finally the core rod is removed.

1.2.1.5.4 Stretch Blow Molding

Stretch blow molding is a modification where the parison is conditioned to suitable temperatures and then biaxially stretched using a stretch rod in the axial direction and air pressure to carry out the radial stretching. In theory, it can be used in both extrusion blow molding and injection blow molding to give specific improved material properties such as transparency, gloss, barrier properties and stiffness, and therefore increased component performance.

The multifunctional applications for injection blow molding need to be embedded at the preform (injection molded stage), but again using coinjection technologies the potential for multifunctional design should not be ignored. A simple schematic of injection blow molding is given in Figure 1.5.

1.2.2 Nonextrusion-Based Processes

1.2.2.1 Thermoforming

Thermoforming is a relatively simple process of forming undertaken by heating a flat preformed sheet and forming it under heat and vacuum onto a male or female mold as required. Because this process uses ready-made plastic film feedstock, there is very little scope for introducing functionality in the process itself, so no further consideration of this process is given here. However, the sheet feedstock can itself contain a number of multilayers and therefore considerable multifunctional potential. Such sheet materials are covered within the sheet and film sections.

1.2.2.2 Lamination

Material film systems that cannot be processed by coextrusion include metals such as aluminum or steel into the structure in packaging applications; there may also be a requirement to use paper-based materials. In these

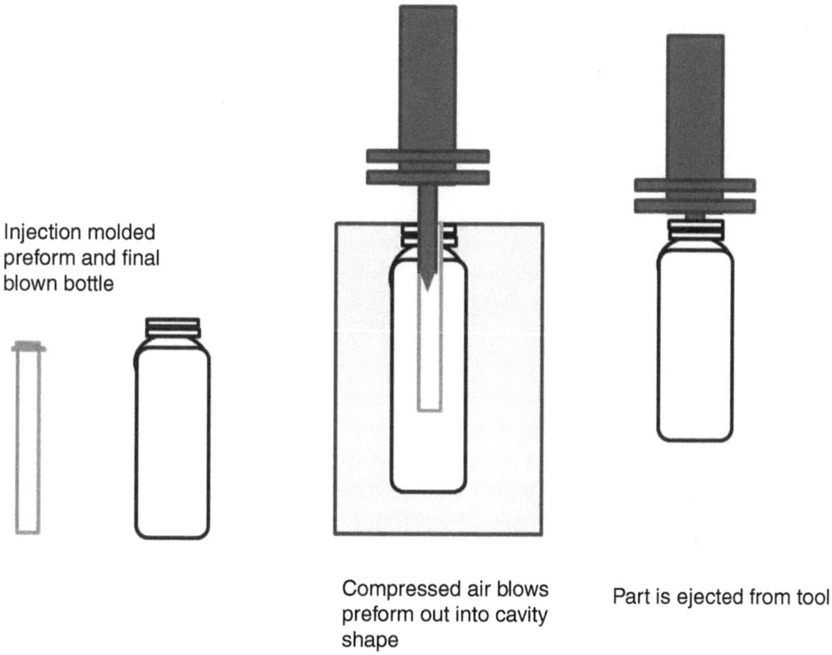

Figure 1.5 Injection blow molding.

cases, lamination and/or coating is needed. The ability to mix materials on a bulk scale is not possible within extrusion screw-based systems (beyond technologies such as wire coating), so lamination presents opportunities of its own, again one embraced by both the food packaging industry (e.g., with food pouches) and battery manufacturers to encapsulate and protect mixed material systems. The lamination process is illustrated in Figure 1.6.

1.2.2.3 Roto-Molding

Roto-molding or rotational molding is a low-pressure process for the production of large hollow objects such as drums. It is an alternative process route to blow molding for the production of hollow objects and is usually limited to the production of fairly large objects such as boat hulls.

The process route is simplistic in that a powder feedstock is placed in the tool cavity which is then rotated and heated. The material falls onto the cavity surface by gravitational force and if heated to melting point will stick to the cavity walls. The distribution is even due to even force and rotation throughout the cavity ensuring consistent wall thicknesses. The part must be cooled before unloading. Multilayered systems can easily be built up using this process, by loading and rotating subsequent materials layers; however,

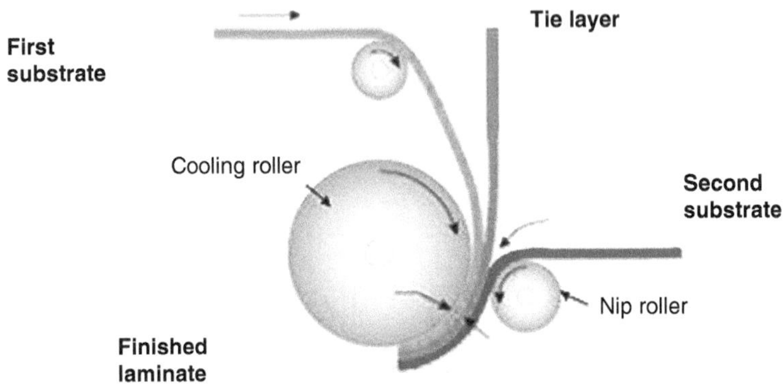

Figure 1.6 Lamination process.

in this case, the necessity of introducing multifunctional design is not clear. In other cases, there are clear potential application areas; for this particular process, it is a niche market and at this point multifunctional application is unclear. It is included here to highlight this point.

1.2.2.4 Low-Volume Composite Manufacture

At the beginning, an example application of a tennis racket with carbon nanotubes was introduced. Sports equipment is an area where advances in material technology can also be seen in the marketplace with extensive use of carbon fiber reinforced materials for tough and durable equipment suitable for both professional and leisure sportspeople. Enhancing already high strength composites by the further addition of materials like CNT and health monitoring for in use performance data is being investigated across a variety of application composite platforms. This is considered further in Chapter 3.

1.3 Limitations

A number of processes have been briefly introduced here and hopefully the potential for multifunctional process design (and therefore this book) is clear. Extrusion-based processes do provide an inherent limitation in themselves, in the need for material and process compatibility, but these processes are tolerant to a wide range of materials and material loadings within any given design structure providing adhesion and therefore structural integrity is considered. However, composite manufacture and 3D printing also offer alternative nonscrew-based manufacturing methods to produce functional designs.

The merits in using one process over another must not be overlooked. An expansive coextrusion-based product could be designed and manufactured, but would lamination have provided a cheaper production solution or could post assembly of a number of separate components produce a more effective final result? Could a combination such as the next example of a multiprocess-based component be the answer? Multifunctional design adds infinite possibilities but also more complex manufacturing decisions to be made in the drive for increasing multipurpose products.

1.4 Example Application

1.4.1 Creating Touch Screen Displays Using In-Mold Labeling (IML)

> A touch screen device produces an input signal in response to a touch or movement of the finger on the display.
> (Helander et al., 1997).

In-mold technology (covered in Chapter 4) enables various processing steps to be combined into one within the mold to achieve added functionality, improved aesthetics, or surface properties. In-mold decoration and IML are common techniques used by manufacturers and designers to improve durability and surface appearance of plastic products. The technique can result in cost savings of up to 40% arising from the removal of secondary operations as well as substantial reductions in carbon emissions (Lentz, 2010).

This application uses a multilayer film produced by a coextrusion process. This film is then integrated into the injection molding process to produce a final assembly. Further complexity can be added to add color, pictures, form, and layer functionality as required.

Applications include electronic devices, automotive, medical equipment, and consumer electronics with increasing interest from product manufacturers who wish to adapt the process for their own uses. There is continuous development of in-mold technology; the majority of these originate from companies working in the packaging, mobile phone, or automotive markets (Goodship and Love, 2002; Love and Goodship, 2002; Chen et al., 2008; Javierre et al., 2014). This makes the process ideal for electronic products such as touch-screen displays commonly used in devices such as game consoles, personal computers, tablet computers, and smartphones (Lentz, 2010, INOTEC, 2012) (Figure 1.7).

This example presents a further aspect of multifunctional design.

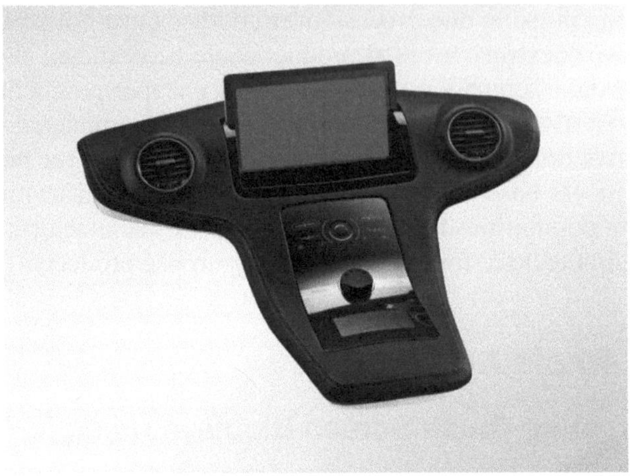

Figure 1.7 Demonstrator for automotive applications with integrated PolyTC® control elements in combination with KURZ IMD films. Image provided by PolyIC GmbH & Co. KG.(PolyIC GmbH & Co. KG, 2014).

1.5 Conclusions

It has been shown that within a plastic design environment, multifunctionality can be introduced in a variety of material level, process level, and component level stages. As advances in material science have led to a number of new high-performance materials, processing technologies have and must continue to develop to take advantage of these developments through integrated, customized, tailored, material level, and component design level solutions. Therefore, the potential in traditional multimaterial extrusion-based processing routes such as coextrusion, coinjection and therefore coextrusion blow molding and multimaterial sheet production have also expanded with the opportunities to combine new and functionalized material combinations and hybrid processes.

For composite manufacture the ability to embed functions within the structure of a component, whether it is a battery on a racing car body panel, or a solar cell embedded on a composite integrated into the building design or a composite able to monitor its own structural integrity all drive research effort forward.

For products such as automotive, telecommunications, and consumer goods the possibilities of producing intelligent technologies for communication, environmental monitoring, and energy production are both exciting and broad. In food packaging active, bioactive, and intelligent production technologies also offer considerable potential for increasing quality of life, be it in health care/monitoring, food preservation and safety, and enhancing the modern environment around us. However, designers are yet to fully utilize the full range of plastics processing available to them, sometimes due to technological limitations, sometimes due to cost but also because of a lack of know-how. It is this final knowledge gap this book will hopefully address to give readers the opportunity to utilize and develop the inherent plastic processing capacity needed to embrace the challenges of new material integration.

References

Arivalagan, K., Ravichandran, S., Rangasamy, K., Karthikeyan, E., 2011. Nanomaterials and its potential applications. Int. J. ChemTech Res. 3, 534–538.

Chen, S.-C., Huang, S.-T., Lin, M.-C., Chien, R.-D., 2008. Study on the thermoforming of PC films used for in-mold decoration. Int. Commun. Heat Mass Transfer 35, 967–973.

Goldberg, H.A., Samantara, L., Rigie, D., Feeney, C.A., Farrell, M., 2011. Moisture-barrier properties of water-based nanocomposite coatings. Soc. Plast. Res. Plast. Res. Online, 1–3, doi:10.2417/spepro.003644.

Goodship, V., Love, J.C., 2002. Multi-material injection moulding. Rapra Technol. 13, 9.

Hayashi, T., Kim, Y.A., Natsuki, T., Endo, M., 2007. Mechanical properties of carbon nanomaterials. ChemPhysChem 8, 999–1004.

Helander, G., Landauer, T.K., Prabhu, P.V., 1997. Handbook of Human–Computer Interaction. Elsevier Science Inc., New York, NY, USA.

INOTEC, 2012. RFID in-mould labels [Online]. Available from: http://www.inotecbsl.com/products/rfid-labels/rfid-in-mould-labels/ (accessed 14.03.2013).

Javierre, C., Fernández, Á., Clavería, I., Elduque, D., 2014. Textile injection process characterisation by means of a spiral mould. Measurement 47, 100–109.

Kampmann, Y., De Clerck, E., Kohn, S., Patchala, D.K., Langerock, R., Kreyenschmidt, J., 2008. Study on the antimicrobial effect of silver-containing inner liners in refrigerators. J. Appl. Microbiol. 104, 1808–1814.

Kirleis, M.A., Simonson, D., Charipar, N.A., Kim, H., Charipar, K.M., Auyeung, R.C.Y., Mathews, S.A., Piqué, A., 2014. Proc. SPIE 8970, Laser 3D Manufacturing, 897004 (March 6, 2014). doi: 10.1117/12.2044222.

Lee, I.-H., 2011. Bonding "Unjoinable" Polymers DuPont Packaging & Industrial Polymers.

Lentz, A.M., 2010. In-mold decorating with pre-printed film inserts. Mod. Plast. Plast. Decorating Mag. 4, 20–23.

Love, J.C., Goodship, V., 2002. In-Mould Decoration of Plastics, vol. 13 (Number 2). Rapra Technology Ltd, Shawbury, UK.

Matic, P., 2003. Overview of multifunctional materials. Smart Struct. Mater. Active materials: behavior and mechanics. Proc. SPIE 5053, vol. 61 (August 12, 2003). doi:10.1117/12.498546.

Tjong, S.C., Liang, G.D., Bao, S.P., 2007. Electrical behavior of polypropylene/ multiwalled carbon nanotube nanocomposites with low percolation threshold. Scrip Mater. 57, 461–464.

2 Materials and Deposition Processes for Multifunctionality

Ruth Cherrington and Jianwang Liang

2.1 Materials

2.1.1 Introduction

Multifunctional materials are designed to perform one or more necessary functions. The most common examples are found in composite or hybrid materials that are combined to perform another function beyond that of structural strength (Thomas and Qidwai, 2004; Leigh, 2010). Electrical and thermal conductivity, magnetic, optical, and many more other functionalities can be integrated into structural materials to provide advantages beyond which traditional materials are capable (Nemat-Nasser et al., 2005). A significant amount of research has been conducted into two-phase structural materials, such as fiber composites, laminates, foams, and other porous structures to provide weight or volumetric benefits. This will be discussed in more detail in Chapter 3.

This chapter will provide an introduction across the broad scope of materials in multifunctional applications. It will look at some deposition processes that can be used to design integrated processes and add-on components, for the benefit of the bulk processes that are the main focus of this book.

These sections can be considered as looking into functions of

Electrical conductivity
Thermal conductivity
Antimicrobial
Magnetic materials
Shape memory

There is an introductory section explaining the main uses of another generic material family: the ceramics. This explores ceramic developments in line with polymeric ones.

Finally, there is a brief section looking specifically at how polymers can be used in sensor applications. The chapter now begins looking at electrical conductivity and how it can be used.

Design and Manufacture of Plastic Components for Multifunctionality
http://dx.doi.org/10.1016/B978-0-323-34061-8.00002-8

2.1.2 Electrical Conductivity

As our world becomes more and more technically advanced and computer technology and electrical devices more prevalent, the ability of materials to conduct electric is a useful property being utilized in a continuing developing range of advanced applications in everything from coatings to embedding function whether as an add-on or within the matrix of a structural component. These are the traditional domains of metals such as copper, but increasingly polymeric-based application solutions are being investigated including applications such as electrostatic discharge, electromagnetic shielding for electronic devices, and even for lightning strike protection.

Electrostatic charges need to be discharged in a controlled way to prevent electrical damage. An uncontrolled electrical discharge may generate high currents that may damage the material or influence the performance of electronic equipment (Cosnier and Karyakin, 2011). For example, static charge that builds up within medical equipment is traditionally discharged by using a metal earthing plate that makes contact with the ground. However, if equipment was encased in a conductive polymer, electric charge could be continually drained away, maintaining a static-free environment (Zhou and Lubineau, 2013). Research into conductive polymers and their development into commercial application has meant demand for materials such as poly(3,4-ethylenedioxythiophene) polystyrene sulfonate (PEDOT:PSS) is on the increase. Therefore, this section will introduce some of the better known conductive polymers after first explaining how materials can conduct electricity.

Conductive materials have a large number of conduction electrons (between one and four electrons per atom). Common examples of high-conductivity materials are metals such as copper, silver, iron, and aluminum. The degree of conductivity is dependent on the number of charge carriers as well as the number of electrons in the material and their mobility. In a metal it is assumed that all the outer electrons are free to carry charge and when a voltage is applied across the metal, the electrons flow from one end of the conductor to the other under the influence of the electric field. Insulating materials such as wood, glass, ceramic, or the vast majority of polymers have tightly bound electrons and thus almost no electron flow activity can be found; as a result, they have high resistance to charge carriers flowing. Their conductivity is very poor. To be a conductor, free electrons need to be present to move. In between metals and insulators, they are materials called semiconductors such as silicon. For these materials, the conductivity can be adjusted by changing temperature or the loading of added charge-carrying agents that are known as dopants. This can be done at a structural or bulk level.

For example, adding charge-carrying materials as dopants (such as metal flakes) to insulating polymers by extrusion a conductive path can be created. There are also additions that can be made at a molecular level that will now be explained in the difference between instrinsically and extrinsically conductive polymers.

2.1.2.1 Intrinsically Conductive Polymers

Conductive polymers are a distinctive group of organic materials that exhibit the electrical and optical properties of both metals and semiconductors (Geoghegan and Hadziioannou, 2013b). They have a wide range of practical applications due to their low cost, conductive ability, and ease of synthesis. More than 25 conductive polymer systems have been reported; however, polyaniline, polyacetylene, polypyrrole, and poly(3,4-ethylenedioxythiophene) (PEDOT) are the most intensively studied.

Polyaniline has been around for hundreds of years and was first reported in 1835 as "aniline black" (Syed and Dinesan, 1991). It is a dark green powder in its conductive form and can be doped to produce high-level conductivities. However, one of the major limitations is that it has poor solubility in commonly known solvents such as sulfuric acid, methane sulfonic acid, formic, and acetic acids. Polyaniline has been successfully used as a coating to prevent the buildup of static energy in electrical components (Aldissi, 1993).

Polyacetylene is a conducting polymer with one of the simplest structures. It was first found to conduct electricity in 1977 and later it was discovered that very heavy doping can produce conductivities similar to that of copper. However, the material was found to be very difficult to work with as it could not be easily dissolved; therefore, other synthesis routes have since been investigated to enable the material to be dissolved into liquid precursors (Townsend et al., 1985). Therefore, uses for bulk plastic materials are unlikely in the near future.

Polypyrrole is a conductive polymer that is commonly used for commercial applications due to its long-term stability. It was first reported in 1916 as "pyrrole black" by the oxidation of pyrrole with hydrogen peroxide to produce an amorphous powder. The applications of polypyrrole were once limited due to its poor solubility; however, recent efforts have enabled the production of soluble polypyrrole formed by graft copolymerization of pyrrole. This has led to applications such as polypyrrole–paint, polypyrrole–polyvinylchloride injection molded composites, and polypyrrole-coated fabrics and fibers (Saville, 2005).

PEDOT was developed by Bayer in the late 1980s as an antistatic coating (Geoghegan and Hadziioannou, 2013b). It has excellent transparency, good electrical conductivity (in excess of 300 s/m) and shows good stability in air and humidity. It has also been found to be stable at relatively high temperatures, with the ability to withstand 125°C for several thousand hours. However, PEDOT is insoluble and therefore very difficult to work with. To enable commercial success PEDOT has been doped with the water-soluble polyelectrolyte, poly (styrenesulfonate) (PSS) resulting in a good conductivity liquid solution known as PEDOT:PSS (Yoshioka and Jabbour, 2006). In addition, it is important to identify the amount of PSS added into PEDOT as it does affect the stability and the quality of the overall dispersion process (Aernouts et al., 2004). With more than 20 years of evolution, PEDOT has become one of the most commercially developed conductive polymers. What is most interesting about the development of PEDOT is it is not the most conducting of the conductive polymers but it is the ability to process it in a variety of ways by doping it that have led to its commercial success. The use of PEDOT-based materials is being heavily researched around the world in a variety of applications from batteries to photovoltaics. It is especially interesting as a replacement for indium tin oxide (ITO) that has wide use as a transparent film-conductive coating in a wide variety of electrical devices but presents manufacturers with supply issues due to its expense.

2.1.2.2 Extrinsically Conductive Polymers

The electrical conductivity of structural polymers can be increased by the addition of conductive fillers, including metals, several forms of carbon, and conductive polymers (Stankovich et al., 2006; Deng et al., 2014; Tiusanen et al., 2012; Mechrez et al., 2012). As well as increasing the electrical conductivity, the inclusion of these fillers can also increase the thermal conductivity and stability, absorb specific radiation, increase the density, and exhibit permanent magnetism (Bhattacharya, 1986). Depending on the type of material, conductive fillers can be used in very small or large quantities to give a plastic the ability to conduct electricity (Mark and Kroschwitz, 2003; Strong, 2006; Dyson, 1997). Glass has been traditionally added to provide strength to a composite; however, the addition of 20–30% glass can increase the conductivity by approximately 1.3 times the normal range for most polymers (Crossman, 1985). To increase the conductivity, glass fibers have been metallized with either aluminum or nickel. However, it is very important with these materials to avoid the breakage of particles that would lead to a dramatic decrease in performance. This may be challenging, if the materials need to withstand the high shear rate environments of extrusion or injection molding, which are processes that many of these conductive fillers were intended for. Glass fibers are further discussed in Chapter 3.

Polymers that incorporate conductive fillers are also known as extrinsically conductive polymers. The level of conductivity will depend on the choice of additive and loading level that can be achieved from the thermoplastic compound. The loading of additives for conductive plastics has a physical impact on polymer properties. Stiffness (flexural modulus) and tensile properties often increase. Impact behavior and elongation, however, generally drop significantly as the type and loading of additive is increased. Benefits to this type of conductive polymer are the relative low cost and its ease of processing (Mark and Kroschwitz, 2003). However, one disadvantage is that properties of the plastic can be negatively affected by the high proportion of filler (Dyson, 1997); further drawbacks including lack of reproducibility stemming from the inhomogenity of the material in the microscale (Mark and Kroschwitz, 2003). Melt mixing techniques such as twin-screw extrusion (Chapter 1) and injection molding (Chapter 4) are the most commonly used manufacturing techniques for conductive polymer composites as they are compatible with current industry methods (Pang et al., 2014).

2.1.2.2.1 Metallic Fillers

Metal fibers and particles including silver, copper, and nickel have been added to plastics to produce conductive composites. They have excellent conductive properties and are largely chemically inert, although oxidation may occur over time.

Thermoplastic resins with a high heat resistance are usually preferred for polymer blends with metallic fillers. Commercial blends often use materials such as polyphenylene oxides, nylon 6/6, elastomeric silicones, polycarbonates, and polyphenylene sulphides.

An important concept for any electrical material system is the percolation threshold. This is the point at which a system goes from having no electrical network to having a connected system. As this is a mathematical concept, numbers will be given in the text for those with prior understanding on this subject. As further discussion is outside the scope of this book, interested readers need to study the concepts of percolation theory.

The amount of material needed can depend on a variety of influences such as the type of particle material (such as copper or aluminum, loading, particle shape that influences the packing density), orientations, and any other coating on the particle that may be added for adhesion or conductive purposes.

As one example silver (Ag) powder was mixed with several polymers including high-density polyethylene (HDPE), polybutylene terephthalate (PBT), and poly (m-xylene adipamide) (MXD6) in Rybak's research

(Rybak et al., 2010). The conductivity of these three increased about 10^{12} times when adding 20 vol% Ag.

The conductivity of metal–polymer composite increases in compression and falls when in extension. It has been reported that these types of conductive polymers exhibit a phenomenon called the nonlinear positive temperature coefficient (PTC) effect, which is where a dramatic change in resistivity is observed upon heating or cooling so that the material switches between a conductor and an insulator (Rybak et al., 2010). The mechanism of the PTC effect is based on the difference of thermal expansion of the metallic filler and the polymer matrix. The thermal expansion at a particular point will lead to the destruction of the continuous conductive network within the composite and the temperature at which this occurs is called commutation temperature (T_{com}) (Yi et al., 2001).

In addition, the performance of PTC effect has been found to be related to factors such as type, size, shape, and hardness of the filler particles, and also the interaction between the filler and polymer matrix (Rybak et al., 2010). Another important aspect that influences the performance of the conductive composite is known as the packing-factor (F). The packing-factor is the highest possible volume fraction of the filler particles in the polymer composite that can be incorporated to produce a continuous network before any voids appear in the material (Mamunya et al., 2002). It is calculated as

$$\frac{V_f}{V_f + V_p} \tag{2.1}$$

in which

V_f is the volume of the filler particles
V_p is the volume of the polymer

The value of packing-factor depends on the size, shape, fraction composition, and formation of the structure of the filling particles. In theory, the statistical F value of monodispersed spherical particles with any size is 0.64 (Mamunya et al., 2002). For the anisotropic fillers, there is a relationship between ratio l/d and packing-factor value. The relationship has an expression of

$$F = \frac{5}{\dfrac{75}{10 + l/d} + l/d}. \tag{2.2}$$

Furthermore, the packing-factor is directly correlated with the percolation threshold (φ_c).

When the value of F is measured, the value of φ_c can be predicted through function

$$\varphi_c = X_c \times F \tag{2.3}$$

where the critical parameter X_c is a constant when there is a lack of interaction between the conductive and nonconductive phases in the polymer composite.

This explains why the size and shape of the filler material have such an influence on the PTC. Carbon nanotubes (CNT) have a very big *l*/*d* ratio, resulting in a small packing-factor value and hence the percolation threshold value is very small, they also have a low PTC intensity due to the strong conductive network. In comparison, metallic spherical particles obtain high PTC intensities, as the conductive network can be easily disconnected as a result of thermal expansion or mechanical stretch (Kim et al., 2004; He et al., 2005).

When a metallic filled polymer composite is expanded, for example, through a raise in temperature, the resistivity increases. Therefore, metallic filled polymer composites have been investigated as current limiting devices through thermal control. Rybak et al. researched the influence of temperature on the properties of conductive polymer composite. Three different polymer matrices (high-density polyethylene (HDPE), polybutylene (PBT), and poly(m-xylene adipamide) (MXD6)) were reinforced with 20% silver fillers and undertook two heating cycles. It was found that the transition from conductor to insulator occurred at different temperatures with different polymer matrices. It was concluded that the intensity of the PTC effect was dependent on the thermal behavior of the polymer. When the heating cycling was repeated, the transfer temperature was slightly shifted to about 5°C to higher. Rybak also investigated the influence of the filler content on the properties of the composite. They found two clear patterns: first, the initial resistivity of the composite decreases as the filler content increases; second, a higher temperature is needed to reach the transition from conductor to insulator for the composites with higher filler content. The higher the filler content in the composite, the smaller the average distance between the particles. As there is more contact between silver particles, it becomes easier from the transfer of electrons.

2.1.2.2.2 Carbon-Based Fillers

Carbon-based materials such as carbon black, carbon fiber, and carbon nanotubes have varied histories of commercial usage. Carbon fiber, for instance, is commercially used in aerospace and aviation, civil engineering, military, car racing, and other competitive sports applications. These materials offer several advantages over metals, including lighter weight and resistance to corrosion. The mechanical strength and flexibility of carbon materials also make them an attractive option for use as fillers in polymers. The topic of carbon fiber as a composite material will not be discussed in this section but more detail on this material and its applications can be found in Chapter 3.

Carbon micro and nanostructures have also been investigated as a replacement for applications where metals would previously have been used for conductive applications; however, conductivity in carbon itself is often significantly lower than in metals (Allaoui et al., 2002; Sandler et al., 1999).

Carbon black has been used as a conductive filler for many decades to decrease the electrical resistivity of insulating polymers. High levels of carbon black can be problematic in plastics as there is an associated reduction in mechanical properties. Significant research was undertaken to investigate ways to decrease the percolation threshold (the point at which the increased loading of filler develops a continuous conductive network is known as the percolation threshold and is specific to each filler material). In order to reduce the amount of carbon black content in the final composite, a number of techniques were investigated including the use of other additives, optimization of the processing conditions, and modification of the particle size distribution and porosity (Verhelst et al., 1977; Geuskens et al., 1991; Geuskens et al., 1987). A lower percolation threshold in the range of 2–3 wt% of carbon black was achieved using these techniques.

The discovery of buckminsterfullerene (C_{60}) in 1985 leads to the discovery of new forms of carbon nanostructures (Geoghegan and Hadziioannou, 2013a). Carbon nanotubes (CNTs), nanosheets (graphene), and nanoballs (fullerene) are all examples of carbon nanostructures with varying levels of conductivity. CNTs are long cylinder tubes consisting of carbon; the material has generated large amounts of research in the field of nanotechnology since the publication of a significant paper by Sumio Iijima in 1991 (Iijima, 1991). In general, it is said to be a lightweight material with strength higher than diamond and electrical and thermal conductivities higher than that of copper. CNTs are known as one of the strongest man-made materials; however, their use in wider applications is still in its infancy. Nanotubes exist in different forms depending on the bond structure, the first two are known as zigzag and armchair, these are illustrated in Figure 2.1. The third form is known as chiral and is characterized by the chiral vector, which defines the structure and properties of the tube (Phillips and Mello, 2006). The structural design has

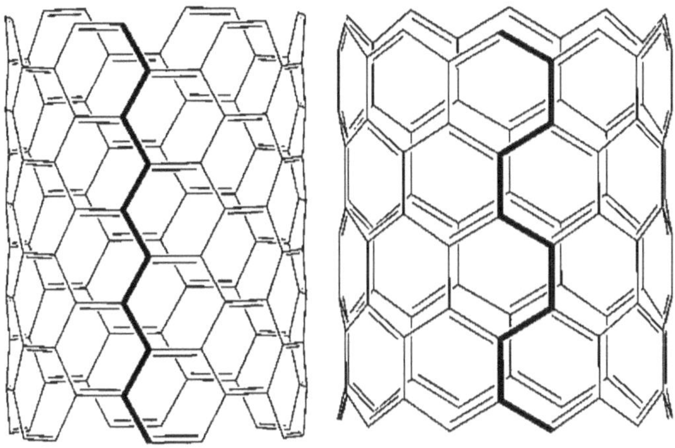

Figure 2.1 Chemical structure of zigzag (left) and armchair (right) carbon nanotubes.

a direct effect on the nanotube's electrical properties. The armchair design is described as "metallic" and is highly conductive, while other designs can make the nanotube a semiconductor.

Nanotubes exist in single (SWCNTs), double (DWCNTs), or multiwalled (MWCTs). Relatively little work has been undertaken on DWCNTs, but it is thought that the DWCNT band structure is unaffected by interlayer interactions (Phillips and Mello, 2006). Their performance also appears to be similar, if not better than SWCNTs. MWCNTs are "metallic" conductors that have more conductive channels per unit volume than the other two forms, which leads to higher charge transport. However, they do suffer from lower optical transmittance because of their large diameters. MWCNTs have physically similar properties to SWCNTs, but the outer walls can protect the inner CNT from damage created from chemical interactions. There are two types of MWCNTs models, one is called Russian Doll model; this type of MWCNT is formed by one SWCNT containing another one inside of it and then another containing the inside two just like the Russian doll (Harris, 2010). Another type of MWCNT model is the rolling up a graphene sheet by several rounds to produce a multilayer carbon nanotube; this type of MWCNT is called the Parchment model.

As previously mentioned, individual CNTs have very high charge carrier properties and current-carrying capacities. Their use in wider applications requires a network of conducting pathways provided by multiple overlapping conductors (Aharony and Stauffer, 2003). Percolation describes the formation of these conducting pathways. For example, consider

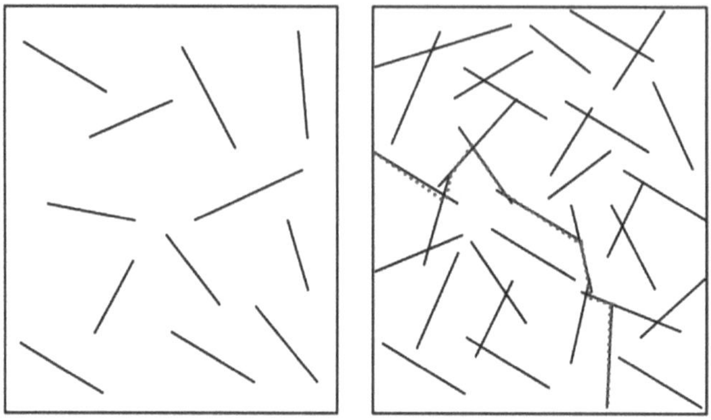

Figure 2.2 Illustration of conducting pathways in a network of CNTs. (a) High sheet resistance due to the lack of a conducting pathway, (b) conducting pathway formed resulting in a drop in sheet resistance.

a thin sheet of material with randomly deposited nanotubes (illustrated in Figure 2.2), when the number (N) of these conductors is small the sheet resistance is high due to the lack of a conducting pathway. As more conductors are added, conducting pathways are formed and the sheet resistance decreases. This point is known as the percolation threshold.

One of the main challenges associated with CNT composites is the creation of a uniform dispersion of particles within the polymer matrix. For example, research investigating the dispersion of multiwalled CNTs in polypropylene found that aggregation of particles forms from the addition of only 0.5 wt% CNTs, with a particle size ranging from 0.2 to 10 μm (Tjong et al., 2007). Larger aggregates are formed as the particle sizes increase and as more CNT content is added to the composite, this is due to the large surface forces that cause the nanotubes to interact with each other.

The potential of carbon nanotubes (CNTs) as a functional filler to increase the electrical conductivity of polymers has been and continues to be an area of significant research. Among the three types of CNTs, armchair-type CNTs are the most conductive model as the transportation of the electrons is along a straight axis. The strong bonds between the carbon atoms enable the CNTs to withstand high currents. The addition of CNTs to polypropylene (PP) results in a decrease of resistivity until it reaches a percolation threshold of 3 wt%. This is illustrated in Figure 2.3, where the electrical resistivity drops dramatically from 10^{13} Ω cm at 0.5 wt% to 10^2 Ω cm at 3 wt% (Szentes et al., 2012).

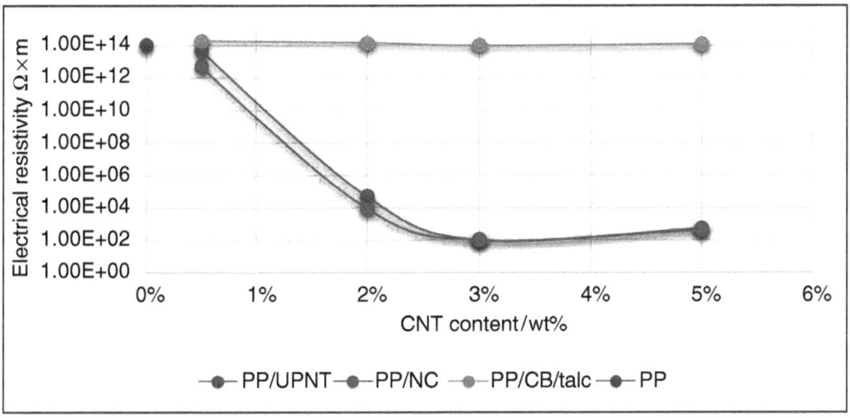

Figure 2.3 Electrical properties of PP-CNT. Adapted from Szentes et al. (2012).

The full potential of CNT to improve the mechanical properties of polymer materials has yet to be fully exploited beyond a loading of around 5%. While addition of CNTs can provide a better increase in mechanical strength combined with weight reduction. Commodity and engineering polymers such as polypropylene (PP), polyethylene terephthalate (PET), and the polyamides continue to be the base materials for a wide variety of thermoplastic polymer nanotube research. For example, the properties of MWNT/PP composite materials have been researched since CNTs were discovered (Prashantha et al., 2009; Gojny et al., 2005; Allaoui et al., 2002; Coleman et al., 2006). However, there are no real property benefits above around 5% loadings for bulk applications. With additions between 0.5 and 5.0 wt%, CNT content increases the tensile modulus (stiffness) but makes the composite more brittle. For this reason, it is the electrical properties rather than the mechanical properties of CNT that tend to be of most commercial application interest.

2.1.2.3 Example Application: Injection Molding to Form Conductive Tracks

As already discussed in this chapter, one way to make polymer melts electrically conductive is through the addition of metal or carbon fillers/additives. However, there is growing demand for the manufacture of components with selectively conductive areas, especially in the field of electronics or automotive.

A new type of hybrid-processing technology called "integrated metal/plastic injection molding" (IMKS) allows the manufacture of electrically conductive plastic/metal components on a single machine in an injection

Figure 2.4 Osram LED manufactured by IMKS.

molding tool without any further processing steps. Conductive tracks can be molded onto plastic carriers using a metal tin alloy that has high electrical conductivities. These materials have a melting temperature in the range of 200–250°C (similar to that of common thermoplastics such as nylon or polybutylene terephthalate) and are already being used for the lead-free soldering in the electronics industry (Hopmann, 2015). This process enables electronic components such as light-emitting devices (LEDs) to be directly integrated into plastic parts. A component is shown in Figure 2.4.

Alternatively, another method reported is through the use of the two-shot process to produce metalized tracks on injection-molded parts. First, a base body is injection molded with a metalizable material forming the shape of the tracks required. The second shot is used to mold a nonmetalizable material around the tracks to form an insulating barrier. Then, the conductor paths are metalized using electroless plating.

2.1.3 Thermally Conductive Materials

Thermally conductive materials are an important class of materials required for a wide range of applications that require heat sinks. These are used in many applications, everything from industrial, medical, automotive, and electrical devices. More specific examples are microchips and lasers in order to control the temperature and dissipate potentially damaging heat rises. Traditionally, these applications have been the domain of metals and ceramics with materials such as aluminum and copper used in this capacity. Other (nonpolymeric) composite materials utilized as heat sink materials are silicon carbide in an aluminum matrix (AlSiC), diamond/copper–silver alloy, and beryllium oxide/beryllium matrix. These can be used in chip manufacture as substrate materials due to similarities in thermal expansion coefficients with silicon semiconductors.

More recently, polymer materials are being commercially used with additive materials in these kinds of applications. There are two distinct functional classes within this subset:

Thermally conductive but electrical insulator
Thermally conductive and electrically conductive

An example of the first is produced through the addition of ceramic additives such as aluminum nitride and boron nitride to a polymer matrix. The use of graphite fibers into a polymer at sufficient loading does both and is therefore useful for RFI shielding in mobile phones for instance.

A wide range of filled polymer materials are now in the thermally conductive marketplace replacing the heavier metal and ceramic materials. This has implications when injection molding such materials for injection molding cooling cycles for instance (see Chapter 4).

2.1.4 Antimicrobial Materials

Incorporation of antimicrobial material has already been commercially applied in several applications including use in household and medical products. There is a continued interest in food applications to prevent spoilage and contamination. For example, the use of antimicrobial nanoparticles in packaging applications has been found to extend the shelf life of fruit and vegetables by decreasing the microbial growth (Azlin-Hasim et al., 2014). There are several ways in which antimicrobial properties can be introduced into a polymer matrix to help control the growth of pathogenic and spoilage microorganisims (Azeredo, 2009). These include the use of inherently antimicrobial polymers, coating, or absorbing of antimicrobials onto the polymer surface and the incorporation of antimicrobial materials (Radheshkumar and Münstedt, 2006). Antimicrobial materials have been investigated by incorporating nanosized particles, which have a higher surface-to-volume ratio when compared to equivalent microsized materials. These can be added to polymers by introducing as a powder additive which is mixed with the polymer before it is melt processes (such as injection molded or extruded). This enables the nanosized particles to be incorporated into the product to produce antimicrobial properties.

Silver is commonly known for its strong toxicity to a wide range of microorganisms. Several factors have been found to influence the microbial

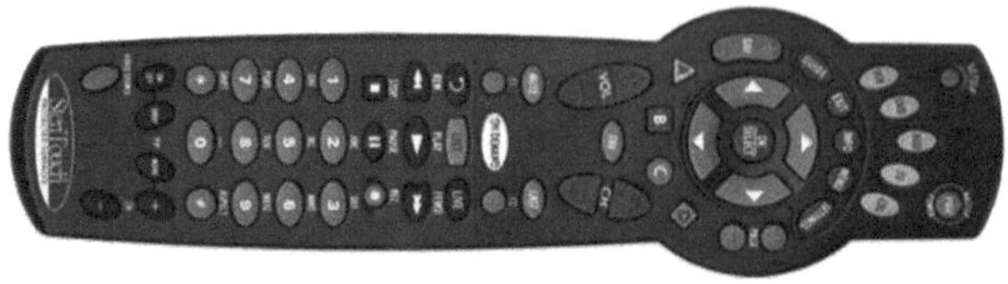

Figure 2.5 Remote control with silver nanoparticles. Image provided with permission from SteriTouch® Ltd. (Corlett, 2014).

growth including the percentage of silver content, the particle size, size distribution, agglomeration, and interaction with the polymer (Azlin-Hasim et al., 2014). A lower silver content has actually been found to have better results on microbial growth; the addition of 2 wt% silver nanoparticles within polyamide-6 has found to be effective against the growth of *E. coli* (Hu and Fu, 2003). Smaller nanoparticles with a narrow size distribution result in better antibacterial activity than larger particles (An et al., 2008). Agglomeration of nanoparticles has been found to significantly decrease the bactericidal effect and research has found that the nanoparticles should be fully dispersed within the polymer matrix in order to fully exploit the antimicrobial effect of silver (Dallas et al., 2011). An interesting application of such technology for common use in a remote control is shown in Figure 2.5.

Titanium dioxide (TiO_2) is a widely used photocatalytic disinfecting material with the ability to reduce bacteria contamination on surfaces (Azeredo, 2009). Titanium dioxide (TiO_2) is a wide bandgap semiconductor, which is transparent in visible light and absorbs a wide ultraviolet (UV) range. TiO_2 has three main crystal phases: anatase, rutile, and brookite. However, the anatase phase shows higher photocatalytic activity than rutile and is the most widely used form (Cerná et al., 2011). When illuminated with UV light, charge carriers are generated that can be used to generate a current, induce chemical reactions, or emit light. This is known as photogenerated catalysis. The photocatalytic activity of TiO_2 increases with decreasing particle size (Brus, 1986) due to an increase in the specific surface area. This is until it reaches a critical particle size of approximately 10 nm, where a loss in photoactivity can be observed (Bertrand et al., 2013). Titanium dioxide is also widely used as a transparent, self-cleaning coating on glass (Paz et al., 1995).

2.1.5 Magnetic Materials

This may be surprising to some, but all material is magnetic to some degree. This is because magnetism is a function of atomic electrons, their spin motions, and how they interact and respond in a magnetic field. What we generally think of as being "magnetic" are those materials that have a strong interaction. These materials are generally classed as ferromagnetic and ferrimagnetic materials though there are two other groups of materials that have weaker interactions called diamagnetic and paramagnet materials.

So in an external magnetic field, a material is influenced as there is an effect on the internal field on that object, that is, the internal fields on its own electrons and atomic nuclei.

These magnetic moments within bulk materials can interact with each other, leading to ordering within the material. This level of ordering results in the different forms of magnetic behavior. If the magnetic moments all line up in parallel then it is said to be ferromagnetic (examples include iron and nickel); if the magnetic moments line up antiparallel with each another then it is said to be antiferromagnetic (an example is chromium). This is illustrated in Figures 2.6 and 2.7.

Examples of diamagnetic material are gold and copper, while platinum and tin produce a paramagnetic effect. Diamagnetic materials exhibit a weak repulsion to a magnetic field and have negative susceptibility. The atomic current loops formed by the orbital motion of electrons respond to oppose the applied magnetic field. Paramagnetic materials have small and positive susceptibility with randomly orientated magnetic moments. Each atom, ion, or molecule of a paramagnetic substance can be considered a tiny magnet with its own intrinsic magnetic moment. These become magnetized when a magnetic field is applied to them. The magnetization depends on the applied field and disappears when the field is removed. The level of magnetism is dependent on temperature. Details of this phenomenon will

Figure 2.6 Illustration of ferromagnetic moments.

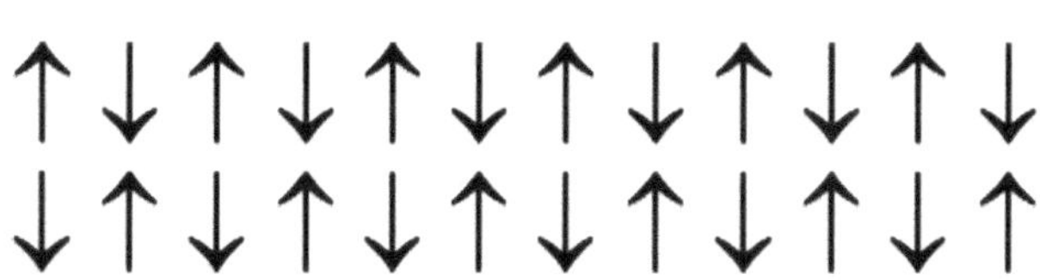

Figure 2.7 Illustration of antiferromagnetic moments.

not be discussed here, but readers interested can refer to the following sources (Coey, 2010; Jiles, 1998; Anderson and Weiss, 1953). Magnetic behavior known as ferrimagnetism is where materials consist of at least two interpenetrating sublattices with antiparallel alignment of the spins. Unlike antiferromagnetic substances, ferrimagnetic substances possess a net magnetic moment and are usually metallic.

Permanent magnets are characterized by their ability to produce a significant magnetic field without the support of an external field. There are currently three main ferrite materials used in magnets in the industry; these are oxides of iron (Fe), barium (Ba), and strontium (Sr). This is mainly due to their low cost. However, more expensive materials with higher magnetic characteristics can be used such as rare earth materials; samarium cobalt (SmCo) magnets, and neodymium iron boron (NdFeB). The level of magnetization of a material depends on how these constituents within a material react with each other. When a magnetic field is applied to the material, a response is generated called its magnetic induction. The amount of magnetic induction created depends on a number of factors, but most importantly, is the magnetization of the material. Small magnetic particles occur naturally in some igneous rocks and small ferromagnetic and ferromagnetic particles can be synthesized by a number of chemical methods.

Magnetic particles can be introduced to a polymer matrix to produce a composite material that shows some response to an applied magnetic field (Ritter et al., 2004). These composite materials have low flexibility and do not change shape, size, or elastic properties in the presence of an external magnetic field and are commonly used to replace steel or ferrites for applications where magnetic properties and minimum weight are required (Shokrollahi and Janghorban, 2007). They can also be formed into a wide range of shapes to produce components for a wide range of applications.

There are several methods that have been developed to synthesize magnetic nanoparticles for the use in composites. More in detail on these can be found in a review by Lu et al. (2007). As with many nanomaterials, there is a problem of maintaining these synthesized nanosize particles without agglomeration. Pure magnetic materials such as iron, cobalt, and nickel are also highly sensitive to air, meaning that it can be difficult to protect them from oxidation in air. The main method to protect the particles is through coating with either organic or inorganic components. Polymers can be chemically anchored or physically adsorbed onto the surface of the particles to prove a single or double layer of protection against the attractive forces acting on the nanoparticles. Polymers successfully used include poly(pyrrole), poly(aniline), and polyesters, such as poly(lactic acid)

(Lu et al., 2007). However, polymer coating does not provide a good enough barrier to prevent against oxidation of very reactive magnetic nanoparticles. More stable coatings have been found through the deposition of precious metals such as platinum or gold. Silica has also been investigated as a coating and found to be stable in acidic solutions.

2.1.5.1 *Ferrites*

A ferrite is a type of ceramic compound composed of iron oxide (Fe2O3) combined chemically with one or more additional metallic elements. They are ferrimagnetic, meaning they can be magnetized or attracted to a magnet, and are electrically nonconductive, one of the few substances that combine these two properties. Ferrites can be divided into two families based on their magnetic coercivity, their resistance to being demagnetized. Hard ferrites have high coercivity; they are difficult to demagnetize. They are used to make magnets, for devices such as refrigerator magnets, loudspeakers, and small electric motors. Soft ferrites have low coercivity. They are used in the electronics industry to make ferrite cores for inductors and transformers, and in various microwave components. Yogoro Kato and Takeshi Takei of the Tokyo Institute of Technology invented ferrite in 1930.

Barium ferrite, abbreviated BaFe, BaM, is the chemical compound with the formula BaFe2O4. This and related ferrite materials are components in magnetic stripe cards and loudspeaker magnets.

2.1.5.2 *Others*

Commercially available magnets include cobalt iron, chromium iron, silicon iron, phosphorous iron, extremely strong bonded neodymium (NdFeB) magnets and sintered aluminum, nickel and cobalt in various proportions (ANiCo) magnets.

Generally in plastics processing metal contamination, including with magnets, is problematic; they are used in hoppers to protect machines from damaging metallic contamination. However, as small particles they can easily be integrated into a polymer matrix to give a greater magnetic property to the polymer.

2.1.6 Shape-Memory Polymers

While there are other classes of shape memory materials such as metals, this section will focus only on shape memory polymers.

Shape memory polymers (SMP) are able to respond to specific stimulus by changing their shape. Most are dual-shape materials that exhibit two distinct shapes: an original and deformed shape. Shape memory polymers should remain in their deformed shape until exposure to the appropriate stimulus; this is when the original shape is recovered. Direct actuation is usually stimulated by heat or light, other stimuli such as magnetic field, electricity, and chemical, for example, pH. Those that are activated by a heat are of particular interest because the temperature at which the shape transition is initiated can be easily controlled. Shape memory polymers are being developed as multifunctional materials so that they not only exhibit the shape-memory effect but also perform a specific function. This section will now continue with a few examples to demonstrate the breadth of applications and industry sectors in which shape-memory polymers can be applied.

Shape memory polyurethanes (SMPUs) are one of the most popular materials to be researched in the field of thermomechanical SMPs due to the ease of their processing and low cost. In the thermomechanical SM cycle, the material is heated above the glass transition temperature (T_g) and mechanical force is required to maintain the deformed shape, when the temperature is cooled below T_g the force can be removed. When the temperature is increased again above T_g, the shape will recover to the initial state. This is called the thermomechanical behavior for SMPs, where the stimulus to recover the shape in this case is heat.

Traditionally, braille writing is made by manually creating raised dots using a stylus and slate or by using a thermoset typing machine to permanently embed dots onto a plastic sheet. Although this method works well for most applications, it is difficult to erase errors when writing. The use of polyurethane SMPs allows for the writer to remove dots by using a pointer heater and then rewrite again. In addition, the whole polyurethane sheet can be heated to restore it back to the original state and hence recycle and use it again This makes it a more sustainable and low-cost alternative to the manufacture if braille writing.

Another potential application of thermomechanical SMPs is for the manufacture of screws for electrical devices and medical applications. Usually, different sized screws fit into different sized holes; however, it is possible to produce a screw that fit into all sized holes with SMP materials. This type of screw can be recovered by heating up and reusing it to manufacture in different size again, resulting in a low waste alternative to manufacturing screws that are usually made of steel.

SMPs are also being applied to modern unmanned aviation vehicles. These require different configurations to achieve different tasks and therefore SMPs can be used to produce wings that reconfigure or "morph" to change shape or size based on flying speed as it can reduce drag remarkably.

However, by far the most popular application of SMPs is in biomedical engineering. Use of materials with a low glass transition temperature at room temperature makes it suitable for biomedical devices, which can be flattened to be implanted at body temperature (37°C) and can then be returned to the desired final shape.

2.1.7 Not Just Polymers! A Brief Introduction to Ceramics

Ceramics make up one of three large classes of solid materials (polymers and metals are the others). Ceramic materials have already been mentioned in several areas of application, but it is worth drawing these together under one heading so the range of available ceramic materials and their application is more obvious to a plastic-based design engineer. As polymer technology develops, metal and ceramic technology is also developing. As many applications (such as electronics) require an integration of materials, it is likely more polymer and ceramic (and metal) hybrid applications will develop in the future.

Ceramics retain their strength at high temperatures (a property known as refractory) and are perhaps best well known for their massive application in the construction industry with materials such as glass, brick, cement, and tile.

Ceramic tiles are used as thermal barriers on NASA's space shuttle program; these protect the frame from temperatures as high as 1600°C.

Ceramic materials are used as insulators, superconductors, semiconductors, and magnets. They are, therefore, a very vital part of the electronics industry.

Transparent silica (glass) fibers have revolutionized telecommunications. These materials at high quality make up fiber optic cables.

Cellular ceramics are used to make catalytic convertors on automotives and can also be used for similar pollution control applications.

As there are different grades of plastics: commodity, engineering, and specialist plastics, there are also advanced ceramic materials. These are not unsurprisingly the materials that have superior properties.

Advanced ceramics can be broken down into the following four subsections:

- structural ceramics: industrial wear parts, bioceramics, cutting tools, and engine components
- electrical and electronic ceramics: capacitors, insulators, integrated circuits packages, substrates, piezoelectrics, magnets, superconductors, and in the future fuel cells

- ceramics coatings: cutting tools, engine components, and high-resistance industrial wear parts
- chemical processing and environmental ceramics: catalysts, catalyst supports, filters, and membrane applications

The ceramic advanced material market is based on powder processing, and therefore nanosized ceramic powders are also being developed along with new processing routes and net shape forming technologies.

The subject of Chapter 4, injection molding and Chapter 5, additive manufacture of two such methods are being investigated along with ceramic methods of gel casting and freezecasting. These methods are being developed to reduce production and finishing costs to make these materials more competitive.

Again, analogous to development in plastic manufacture there are continuous research and developments in ceramic matrix composites (CMCs) and "smart ceramics" to enhance control by sensing and actuating. There are growing medical applications for replacement of bone and teeth, while ceramic-based superconducting magnets are used in magnetic levitation (maglev) trains.

While this section is just a glimpse into ceramic technology, it is hoped this information may help develop ideas for further hybrid polymer/ceramic applications.

2.2 Film-Forming Processes

2.2.1 Introduction

As discussed in the introduction chapter, one of the methods to provide additional performance to a system is through a simple coating or lamination process. A film-forming process is usually required to facilitate this; therefore, this section reviews some of the techniques suitable for the formation of film.

Many of the techniques require large amounts of material and are difficult to reproduce or are unsuitable for large-scale production. Two examples are spin-coating (Figure 2.8) and electro-spinning, which are both widely used within the research community for thin film deposition, and however are only suitable for batch manufacture. Spin coating is a standard method used to produce uniform coatings of standard thickness. The technique has been used for years in the microelectronics industry to provide a highly reproducible film to substrate with diameters of up to 30 cm (Krebs, 2009). The operation involves application of a liquid to a rotating

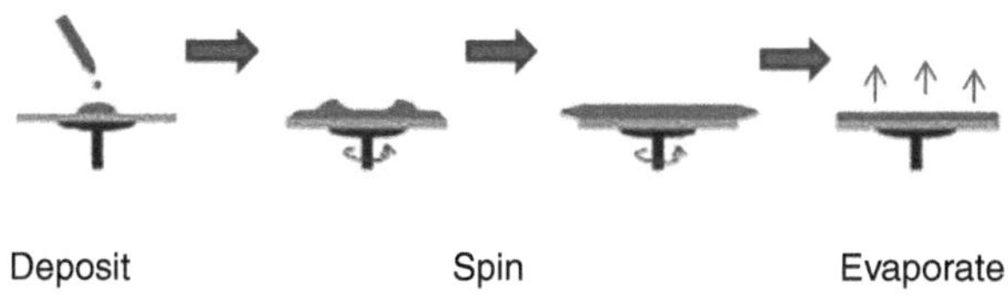

Figure 2.8 Principle of the spincoating process (Reich, 2015).

substrate, altering the spinning rate and time to achieve different thicknesses, film morphology, or surface topography. However, most of the applied solution is wasted during rotation. Spin coating is a very useful approach for laboratory device optimization but it is unlikely to be used for large-scale, high-volume manufacture. There are several reasons for this:

1. Substrates need to be handled individually; therefore, the process is not suitable for large-scale processing.
2. The process does not enable patterning that is desirable for applications such as current collecting grids within cells.
3. The high volume of material wastage of more than 90% is likely to result in high manufacturing costs (Krebs, 2009).

The electrospinning process pushes a polymer solution through a jet by applying a voltage; the voltage difference is about 10 kV between the metal opening and the collector. In addition, this process produces very thin fiber with diameter of 0.5–5 μm (Doshi and Reneker, 1993). The process can be used to produce nanofibers and fabrics with controllable pore structure. The scale up of nanofibers through single jet is not very feasible and therefore several research groups have investigated the use of multiple jets to increase the production rate. The process is thought to have a wide range of applications but is intensively researched in the medical sector (Bhardwaj and Kundu, 2010).

The ideal film-forming process should allow high-speed coating or printing at very low process costs. It should provide solution processing of all layers on flexible substrates with as few steps as possible. It is generally accepted that roll-to-roll (RTR) processing is cost effective and efficient and methods compatible with RTR are mainly explored in the current literature (Na et al., 2010; Angmo et al., 2012; Galagan et al., 2011; Krebs et al., 2009; Larsen-Olsen et al., 2012; Park et al., 2010; Søndergaard et al., 2012; Steim et al., 2011; Krebs et al., 2010). RTR methods can be prepared without vacuum, using water solvents and can be used to process a wide range of flexible substrate materials in seconds (Søndergaard et al., 2012). RTR processing is a processing technique that involves a substrate that is unwound from a roll to allow for continuous printing or coating. Several treatment steps (including heating, drying, UV curing, etc.) and further layers can be added as required (Krebs, 2009). Deposition techniques can be separated into two categories, the first of which is vacuum deposition. This is a high-energy, high-cost process that is performed at high temperatures. The second is that of processes that utilize solution-based materials; these printing and coating techniques can be undertaken in atmospheric conditions and relatively lower cost to that of vacuum processes.

There are a wide range of materials that can be formulated into a suitable form for printing. This is usually achieved by either dissolving or dispersing nanomaterials into a solvent carrier to form an ink. Conductive materials have been widely investigated within the field of printed electronics and be separated into three main categories: metallic colloids, carbon nanostructures, and conductive polymers. A summary of the key film-forming techniques compatible with RTR processing will now be discussed.

2.2.2 Vacuum Deposition

Vacuum deposition is a well-known processing technique used in industry to deposit film layers with thicknesses between a few nanometers to about 100 micrometers. The coatings are most often reflective metals such as the deposition of aluminum onto a polymer film for the use in crisp packets. Layers are deposited atom by atom or molecule by molecule by evaporating onto a solid surface at pressures well below atmospheric pressure and therefore require large amount of energy to produce these conditions.

2.2.3 Screen Printing

In screen printing, a stencil is applied to a woven material (synthetic fiber or steel mesh) that is stretched over a frame. The stencil blocks all openings except the patterned area and as a result, there should be no loss of coating solution during printing. The pattern is obtained by applying the coating solution to one end of the frame and (using a squeegee) is pulled across the screen forcing it through the patterned areas and onto the substrate. The thickness of the coating is determined by the thickness of the solution and the screen as shown in Figure 2.9 (Krebs, 2009).

2.2.4 Doctor Blading

Doctor blade coating is a technique used to form films with well-defined thicknesses. The technique works by placing a sharp blade at fixed

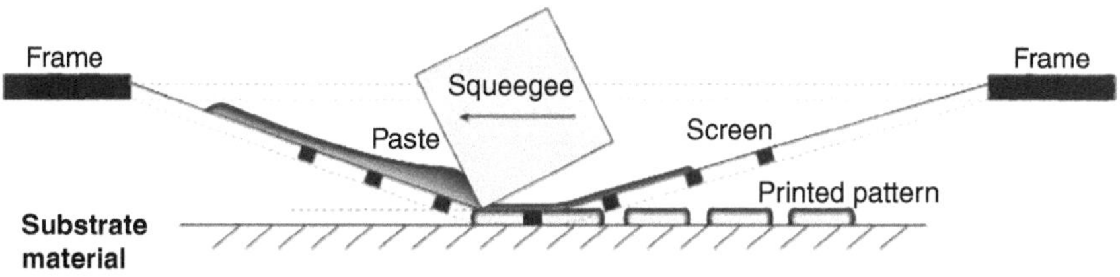

Figure 2.9 Illustration of screen printing process.

distance from the surface that needs to be covered. The coating solution is then placed in front of the blade and the blade is moved across in-line with the surface, creating a wet film. The technique should ideally have solution losses of about 5%; however, practically, it takes time for optimal conditions to be found (Krebs, 2009).

The inks/pastes used in these processes usually require large amounts of binders and thickeners to produce the high viscosities (1000–10,000 mPa s) required for reproducible and reliable production of films (Hoth et al., 2013). Viscosities can be increased with the addition of polymeric additives such as glycerol or ethylene glycol or ethyl cellulose (Di Risio and Yan, 2007).

2.2.5 Slot-Die Coating

Slot-die coating is a simple technique where the coating liquid is transferred to the substrate by being forced out of a reservoir through a slot by pressure. Generally, the reservoir is significantly larger than the slot and is aligned perpendicular to the substrate; however, slot-die coating has several variations to include different die designs and processing of numerous materials.

2.2.6 Gravure Coating

Gravure coating enables direct pattern coating by using a gravure roll: a cylinder engraved with grooves or cells (Yin and Kumar, 2006). The engraved roller runs into a coating bath that fills the engraved parts of the roller with the coating liquid. Excess liquid is removed with a doctor blade and the pattern is deposited onto the substrate as it passes between the etched roller and a pressure roller. The amount of liquid transferred will affect the thickness and print resolution of the coating (Krebs, 2009).

2.2.7 Spray Coating

Spray coating is a technique where the coating material is forced through a nozzle to produce a fine aerosol (Figure 2.10). The process can

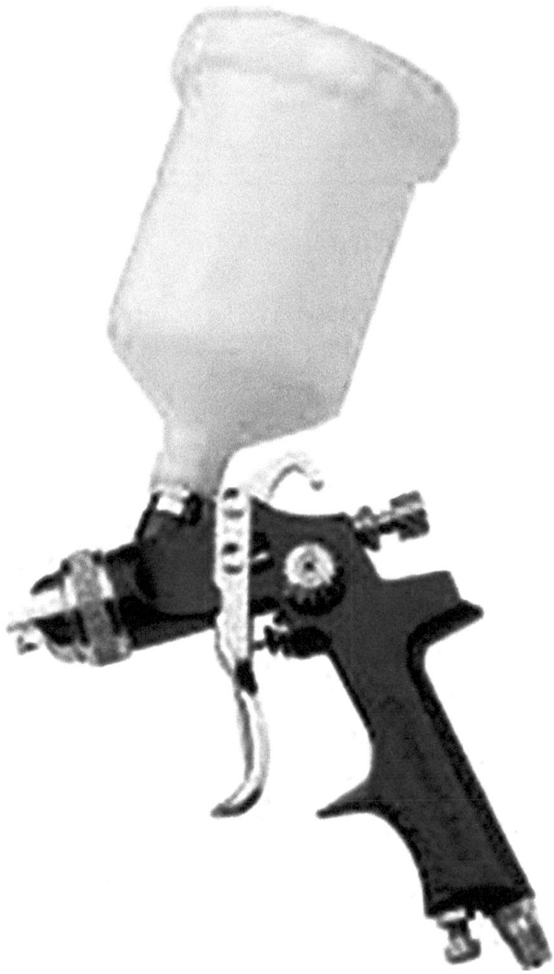

Figure 2.10 Gravity-fed high pressure spray gun.

be combined with a carrier gas and electrostatic charging to improve the direction and quality of deposition of the aerosol spray. However, it can be difficult to produce a smooth film as aerosol droplets do not always level when they reach the substrate surface (Kang et al., 2012).

Spray coating is a well-established industrial technique that is capable of producing large area, uniform RTR coatings with minimal solvent waste (Kumar et al., 2012). The method can be used to produce multilayer electronic devices, and several studies have investigated the technique as an alternative to spin coating resulting in similar performances (Chen et al., 2010; Green et al., 2008; Kang et al., 2012; Kim et al., 2012;

Kumar et al., 2012; Na et al., 2010; Park et al., 2011; Steirer et al., 2009; Susanna et al., 2011). Spray coating is used in a wide range of industries and applications. There are over 90,000 nozzle designs that can be employed to produce the desired spray pattern and angle.

2.2.8 Inkjet Printing

Inkjet technologies can be described as the "digitally controlled ejection of drops of fluid from a print head onto a substrate" (Magdassi, 2010). The technique also implies that there is some form of patterning involved. Whereas, coating is used to describe a technique by which a layer of ink is transferred to a substrate to form a covering. Inkjet printing is unique in that it is a noncontact method that can be classified as both a coating and printing technique. Ink is dispensed directly from a reservoir to a substrate in a predefined pattern (Krebs, 2009).

Most systems operate by dispensing droplets one at a time as the print head moves in the X direction and the substrate moves in the Y. More advanced systems also incorporate a Z movement. It is a noncontact method that can be used on a wide range of substrates including flexible and three-dimensional geometries. Since inkjet printing uses liquid forms of materials dispersed in a solvent, it is critical to understand the deposition of ink to determine the applications of the technique.

The technology is broadly divided into two categories, continuous (CIJ) and drop-on-demand (DOD), with the main difference evident in their names. CIJ printers produce a continuous stream of droplets from the nozzles, whereas DOD printers produce droplets when prompted to do so. There are several similarities between the two methods. First, they both use a pump to provide suction and to prevent ink leaving the nozzles when they are not in use; it can also be used to provide pressure to flush the nozzles with cleaning fluid. They both use an electric motor to correctly position the nozzles above the substrate to produce a pattern. This is either done by moving the nozzles (print head) themselves, the substrate, or both.

2.2.8.1 Continuous (CIJ)

CIJ systems produce a continuous stream of droplets from a nozzle that is directly connected to a reservoir. As the stream of droplets falls toward the substrate, they pass between two electrostatic charging plates. When required, a selection of these electrostatically charged droplets can be deflected, collected in a gutter, and recycled. The technology enables a high drop frequency and also has the ability to use volatile solvents, which allows for a faster drying and increased adhesion to substrates. The process also enables printing at large distances from the substrate due to the high drop velocity. These points combined make possible high-speed printing for applications such as data coding of products and are most commonly used in industry to produce expiry dates on pots and bottles (Cummins and Desmulliez, 2012). However, CIJ results in a relatively low print resolution and has limitations due to the requirement for the print fluid to be electrically charged. The technique is also known for requiring high maintenance and is perceived to be environmentally unfriendly due to the solvents used (Magdassi, 2010). CIJ systems are not commonly used to manufacture printed electronics due to the fact that the quality of the functional inks is compromised when exposed to environmental conditions during the ink recycling process (Angelo, 2013).

2.2.8.2 *Drop-on-Demand*

Thermal inkjet printers operate by rapidly heating the ink inside a reservoir using a resistor connected to a power supply (Sridhar et al., 2011). A film of ink is developed directly above the heater, which vaporizes producing an air bubble. The resultant air bubble causes a pressure pulse, which forces a drop of ink through the nozzle (Angelo, 2013). Thermal inkjet printing has the potential for small drop sizes, high nozzle density, and low print head costs. However, the heating element can lead to degradation of the functional material and is limited in the range of solvents that can be used (Yin et al., 2010).

Piezoelectric printers contain a piezoelectric crystal within the ink reservoir, which changes shape when a voltage is applied (Cummins and Desmulliez, 2012). This results in a pressure wave within the ink reservoir, forcing out a droplet from the nozzle (Angelo, 2013). This piezo-element means that the resultant print head is significantly more costly than the thermal print heads. The drop size and shape can be carefully controlled by adjusting the waveform, meaning that a range of inks and materials can be printed to the pattern required. Smaller droplets can produce patterns of high resolution, whereas larger droplets can be used to cover larger areas and therefore increase the print. Adjustment of the peak voltage, voltage pulse time, and the shape of the waveform applied to the piezo-element can affect how the droplets are deposited (Cummins and Desmulliez, 2012; Sridhar et al., 2011; Tekin et al., 2008; Yin et al., 2010).

Piezoelectric printers are most widely known for their use in desktop printing within the home. As anyone will know who has owned an inkjet printer, one of the main disadvantages is that the narrow inkjet nozzles are prone to clogging. This has largely been overcome through improved ink formulations and automated cleaning cycles. However, it is advised that the print heads are in continual use to avoid ink from drying in the nozzles that may cause blockages. Functional fluids can be formulated for inkjet printing to provide predictable and repeatable fluid deposition with a high resolution. Many of these materials can be expensive, but inkjet technology avoids wastage by only delivering the required amount of material for each application. The technology is flexible which means that it can be scaled up for high output production, and the noncontact nature of inkjet enables it to be used on a wide range of substrates without my risk of damage.

2.3 Sensors

Having looked at various polymer materials deposition techniques, a further kind of polymer material system will now be introduced. The use of polymer film materials for sensors is widespread being a low-cost and relatively easy deposition method (spinning, casting, screen printing, vacuum deposition are all used to produce polymeric films for sensors) for production. In conjunction along the ability to tailor polymer materials a wide range of sensing structures can be produced. These include

- Impedance
- Resonant
- Electrochemical cells
- Calorimetric sensors
- Fiber optic sensors

Application of sensing polymers can allow measurement of temperature, mechanical actuation (i.e., touch, pressure, deformation), electrical actuation (piezopolymers, an example is PVDF), biosensors (everything from enzymes to material identification).

2.4 Conclusions

This chapter has attempted to cover a very broad range of materials and some deposition processes that can be potentially linked to or utilized in bulk manufacturing plastic processes. While this chapter does not attempt or claim to be exhaustive, it is meant to be linked to the more intensive process Chapters 4, 5, and 6 with the intention to give a designer or engineer a broad spectrum of ideas and opportunities from which to gain inspiration. For example, in many ways, inkject printing is the next development and refinement in the growing development of additive manufacturing. Introducing ceramics serves not only to potentially develop mixed hybrid systems, but also to remind the reader that development does not rest in the polymer field alone in advancing manufacturing processes. The multifunction materials here can be used for a variety of purposes to build intelligence through embedding both electrical functionality with a sensing and acting capability into a manufacturing system.

References

Aernouts, T., Vanlaeke, P., Geens, W., Poortmans, J., Heremans, P., Borghs, S., Mertens, R., Andriessen, R., Leenders, L., 2004. Printable anodes for flexible organic solar cell modules. Thin Solid Films 451, 22–25.

Aharony, A., Stauffer, D., 2003. Introduction to Percolation Theory. Taylor and Francis, London.

Aldissi, M., 1993. Intrinsically Conducting Polymers: An Emerging Technology. Springer, Berlin, Germany.

Allaoui, A., Bai, S., Cheng, H.M., Bai, J.B., 2002. Mechanical and electrical properties of a MWNT/epoxy composite. Compos. Sci. Technol. 62, 1993–1998.

An, J., Zhang, M., Wang, S., Tang, J., 2008. Physical, chemical and microbiological changes in stored green asparagus spears as affected by coating of silver nanoparticles-PVP. LWT Food Sci. Technol. 41, 1100–1107.

Anderson, P.W., Weiss, P.R., 1953. Exchange narrowing in paramagnetic resonance. Rev. Mod. Phys. 25, 269–276.

Angelo, P.D., 2013. Inkjet-printed light-emitting devices: applying inkjet microfabrication to multilayer electronics. Doctor of Philosophy, Univeristy of Toronto. Toronto, ON.

Angmo, D., Hösel, M., Krebs, F.C., 2012. All solution processing of ITO-free organic solar cell modules directly on barrier foil. Solar Energy Mater. Solar Cells 107, 329–336.

Azeredo, H.M.C.D., 2009. Nanocomposites for food packaging applications. Food Res. Int. 42, 1240–1253.

Azlin-Hasim, S., Cruz-Romero, M.C., Ghoshal, T., Morris, M.A., Cummins, E., Kerry, J.P., 2014. Application of silver nanodots for potential use in antimicrobial packaging applications. Innovative Food Sci. Emerging Technol. 27, 136–143.

Bertrand, F., German, S.-A., Anwar, A., Irune, V., Gemma, B., Yolanda, R.D.M., Lennart, B., 2013. Dispersion and surface functionalization of oxide nanoparticles for transparent photocatalytic and UV-protecting coatings and sunscreens. Sci. Technol. Adv. Mater. 14, 023001.

Bhardwaj, N., Kundu, S.C., 2010. Electrospinning: a fascinating fiber fabrication technique. Biotechnol. Adv. 28, 325–347.

Bhattacharya, S.K., 1986. Metal Filled Polymers. Taylor and Francis, London.

Brus, L., 1986. Electronic wave functions in semiconductor clusters: experiment and theory. J. Phys. Chem. 90, 2555–2560.

Cerná, M., Veselý, M., Dzik, P., 2011. Physical and chemical properties of titanium dioxide printed layers. Catal. Today 161, 97–104.

Chen, L.-M., Hong, Z., Kwan, W.L., Lu, C.-H., Lai, Y.-F., Lei, B., Liu, C.-P., Yang, Y., 2010. Multi-source/component spray coating for polymer solar cells. ACS Nano 4, 4744–4752.

Coey, J.M.D., 2010. Magnetism and Magnetic Materials. Cambridge University Press, Cambridge.

Coleman, J.N., Khan, U., Blau, W.J., Gun'ko, Y.K., 2006. Small but strong: a review of the mechanical properties of carbon nanotube–polymer composites. Carbon 44, 1624–1652.

Corlett, N. 2014. Remote control SteriTouch® Ltd.

Cosnier, S., Karyakin, A., 2011. Electropolymerization: concepts. Materials and ApplicationsWiley, New York.

Crossman, R.A., 1985. Conductive composites past, present, and future. Polym. Eng. Sci. 25, 507–513.

Cummins, G., Desmulliez, M.P., 2012. Inkjet printing of conductive materials: a review. Circuit World 38, 193–213.

Dallas, P., Sharma, V.K., Zboril, R., 2011. Silver polymeric nanocomposites as advanced antimicrobial agents: classification, synthetic paths, applications, and perspectives. Adv. Colloid Interface Sci. 166, 119–135.

Deng, H., Lin, L., Ji, M., Zhang, S., Yang, M., Fu, Q., 2014. Progress on the morphological control of conductive network in conductive polymer composites and the use as electroactive multifunctional materials. Prog. Polym. Sci. 39, 627–655.

Di Risio, S., Yan, N., 2007. Piezoelectric ink-jet printing of horseradish peroxidase: effect of ink viscosity modifiers on activity. Macromol. Rapid Commun. 28, 1934–1940.

Doshi, J., Reneker, D.H. Electrospinning process and applications of electrospun fibers. In: Industry Applications Society Annual Meeting, 1993, Conference Record of the 1993 IEEE, 1993, pp. 1698–1703.

Dyson, R.W., 1997. Specialty Polymers. Springer, The Netherlands.

Galagan, Y., De Vries, I.G., Langen, A.P., Andriessen, R., Verhees, W.J.H., Veenstra, S.C., Kroon, J.M., 2011. Technology development for roll-to-roll production of organic photovoltaics. Chem. Eng. Process. Process Intensification 50, 454–461.

Geoghegan, M., Hadziioannou, G., 2013a. Beyond polyacetylene. Polymer Electronics. OUP Oxford, Oxford.

Geoghegan, M., Hadziioannou, G., 2013b. Polymer Electronics. OUP Oxford, Oxford.

Geuskens, G., De Kezel, E., Blacher, S., Brouers, F., 1991. The electrical conductivity of polymer blends filled with carbon black—II. Eur. Polym. J. 27, 1261–1264.

Geuskens, G., Gielens, J.L., Geshef, D., Deltour, R., 1987. The electrical conductivity of polymer blends filled with carbon-black. Eur. Polym. J. 23, 993–995.

Gojny, F.H., Wichmann, M.H.G., Fiedler, B., Schulte, K., 2005. Influence of different carbon nanotubes on the mechanical properties of epoxy matrix composites – a comparative study. Compos. Sci. Technol. 65, 2300–2313.

Green, R., Morfa, A., Ferguson, A.J., Kopidakis, N., Rumbles, G., Shaheen, S.E., 2008. Performance of bulk heterojunction photovoltaic devices prepared by airbrush spray deposition. Appl. Phys. Lett. 92, 033301.

Harris, P. 2010. Carbon nanotube science and technology. Available from: http://www.personal.reading.ac.uk/~scsharip/tubes.htm (accessed 17.04.2015).

He, X.J., Du, J.H., Ying, Z., Cheng, H.M., 2005. Positive temperature coefficient effect in multiwalled carbon nanotube/high-density polyethylene composites. Appl. Phys. Lett. 86, 062112.

Hopmann, C., Bobzin, K., Weber, M., Öte, M., Ochotta, P., Liao, X., 2015. IMKS and IMMS—Two integrated methods for the one-step-production of plastic/metal hybrid parts. In: Brecher, C. (Ed.), Advances in Production Technology. Springer International Publishing. http://link.springer.com/book/10.1007%2F978-3-319-12304-2

Hoth, C.N., Schilinsky, P., Choulis, S.A., Balasubramanian, S., Brabec, C.J., 2013. Solution-processed organic photovoltaics. In: Cantatore, E. (Ed.), Applications of Organic and Printed Electronics, Integrated Circuits and Systems. Wiley-VCH Verlag, Berlin, Germany.

Hu, A.W., Fu, Z.H., 2003. Nanotechnology and its application in packaging and packaging machinery. Packag. Eng. 24, 22–24.

Iijima, S., 1991. Helical microtubules of graphitic carbon. Nature 354, 56–58.

Jiles, D.C., 1998. Introduction to Magnetism and Magnetic Materials, second ed. Taylor and Francis, London.

Kang, J.-W., Kang, Y.-J., Jung, S., Song, M., Kim, D.-G., Su Kim, C., Kim, S.H., 2012. Fully spray-coated inverted organic solar cells. Solar Energy Mater. Solar Cells 103, 76–79.

Kim, J.-I., Kang, P.H., Nho, Y.C., 2004. Positive temperature coefficient behavior of polymer composites having a high melting temperature. J. Appl. Polym. Sci. 92, 394–401.

Kim, Y., Kim, G., Lee, J., Lee, K., 2012. Morphology controlled bulk-heterojunction layers of fully electro-spray coated organic solar cells. Solar Energy Mater. Solar Cells 105, 272–279.

Krebs, F.C., 2009. Fabrication and processing of polymer solar cells: a review of printing and coating techniques. Solar Energy Mater. Solar Cells 93, 394–412.

Krebs, F.C., Gevorgyan, S.A., Alstrup, J., 2009. A roll-to-roll process to flexible polymer solar cells: model studies, manufacture and operational stability studies. J. Mater. Chem. 19, 5442–5451.

Krebs, F.C., Tromholt, T., Jorgensen, M., 2010. Upscaling of polymer solar cell fabrication using full roll-to-roll processing. Nanoscale 2, 873–886.

Kumar, P., Shin, P., Ochiai, S., 2012. Fabrication of PTB7:PC71BM bulk hetero junction polymer solar cells by airbrush spray-coating technique. In: 2012 19th International Workshop on Active-Matrix Flatpanel Displays and Devices (AM-FPD), July 4–6, 2012, pp. 293–296.

Larsen-Olsen, T.T., Andreasen, B., Andersen, T.R., Böttiger, A.P.L., Bundgaard, E., Norrman, K., Andreasen, J.W., Jørgensen, M., Krebs, F.C., 2012. Simultaneous multilayer formation of the polymer solar cell stack using roll-to-roll double slot-die coating from water. Solar Energy Mater. Solar Cells 97, 22–27.
Leigh, S., 2010. Smart and functional materials in the future of micro manufacturing. Commercial Micro Manuf. Mag. 3, 19–22.
Lu, A.-H., Salabas, E.L., Schüth, F., 2007. Magnetic nanoparticles: synthesis, protection, functionalization, and application. Angew. Chem. Int. Ed. 46, 1222–1244.
Magdassi, S., 2010. The Chemistry of Inkjet Inks. World Scientific, Singapore.
Mamunya, Y.P., Davydenko, V.V., Pissis, P., Lebedev, E.V., 2002. Electrical and thermal conductivity of polymers filled with metal powders. Eur. Polym. J. 38, 1887–1897.
Mark, H.F., Kroschwitz, J.I., 2003. Encyclopedia of Polymer Science and Technology. Wiley, New York.
Mechrez, G., Suckeveriene, R.Y., Zelikman, E., Rosen, J., Ariel-Sternberg, N., Cohen, R., Narkis, M., Segal, E., 2012. Highly-tunable polymer/carbon nanotubes systems: preserving dispersion architecture in solid composites via rapid microfiltration. ACS Macro Lett. 1, 848–852.
Na, S.-I., Yu, B.-K., Kim, S.-S., Vak, D., Kim, T.-S., Yeo, J.-S., Kim, D.-Y., 2010. Fully spray-coated ITO-free organic solar cells for low-cost power generation. Solar Energy Mater. Solar Cells 94, 1333–1337.
Nemat-Nasser, S., Nemat-Nasser, S., Plaisted, T., Starr, A., Amirkhizi, A.V., 2005. Multifunctional materials. In: Bar-Cohen, Y. (Ed.), Biologically Inspired Technologies. CRC Press, Taylor & Francis Group, Boca Raton, Florida, United States.
Pang, H., Xu, L., Yan, D.-X., Li, Z.-M., 2014. Conductive polymer composites with segregated structures. Prog. Polym. Sci. 39(11), 1908–1933.
Park, H.J., Kang, M.-G., Ahn, S.H., Guo, L.J., 2010. A facile route to polymer solar cells with optimum morphology readily applicable to a roll-to-roll process without sacrificing high device performances. Adv. Mater. 22, E247–E253.
Park, S.-Y., Kang, Y.-J., Lee, S., Kim, D.-G., Kim, J.-K., Kim, J.H., Kang, J.-W., 2011. Spray-coated organic solar cells with large-area of 12.25 cm^2. Solar Energy Mater. Solar Cells 95, 852–855.
Paz, Y., Luo, Z., Rabenberg, L., Heller, A., 1995. Photooxidative self-cleaning transparent titanium dioxide films on glass. J. Mater. Res. 10, 2842–2848.
Phillips, T.W., Mello, J.C.D., 2006. New materials for transparent electrodes. In: Klauk, H. (Ed.), Organic Electronics: Materials, M., and Applications. Wiley, New York.
Prashantha, K., Soulestin, J., Lacrampe, M.F., Krawczak, P., Dupin, G., Claes, M., 2009. Masterbatch-based multi-walled carbon nanotube filled polypropylene nanocomposites: assessment of rheological and mechanical properties. Compos. Sci. Technol. 69, 1756–1763.
Radheshkumar, C., Münstedt, H., 2006. Antimicrobial polymers from polypropylene/silver composites—Ag+ release measured by anode stripping voltammetry. Reactive Funct. Polym. 66, 780–788.

Reich, S. 2015. Principle of the spincoating process [Online]. Available from: http://commons.wikimedia.org/wiki/File%3ASpincoating.svg (accessed 17.04.2015).

Ritter, J.A., Ebner, A.D., Daniel, K.D., Stewart, K.L., 2004. Application of high gradient magnetic separation principles to magnetic drug targeting. J. Magn. Magn. Mater. 280, 184–201.

Rybak, A., Boiteux, G., Melis, F., Seytre, G., 2010. Conductive polymer composites based on mettallic nanofiller as smart materials for current limiting devices. Compos. Sci. Technol. 70, 410–416.

Sandler, J., Shaffer, M.S.P., Prasse, T., Bauhofer, W., Schulte, K., Windle, A.H., 1999. Development of a dispersion process for carbon nanotubes in an epoxy matrix and the resulting electrical properties. Polymer 40, 5967–5971.

Saville, P., 2005. Polypyrrole: formation and use. Defence Res. Dev. Canada.

Shokrollahi, H., Janghorban, K., 2007. Soft magnetic composite materials (SMCs). J. Mater. Process. Technol. 189, 1–12.

Søndergaard, R., Hösel, M., Angmo, D., Larsen-Olsen, T.T., Krebs, F.C., 2012. Roll-to-roll fabrication of polymer solar cells. Materials Today 15, 36–49.

Sridhar, A.B., Blaudeck, T., Baumann, R.R., 2011. Inkjet printing as a key enabling technology for printed electronics. Mater. Matters 6, 12–15.

Stankovich, S., Dikin, D.A., Dommett, G.H.B., Kohlhaas, K.M., Zimney, E.J., Stach, E.A., Piner, R.D., Nguyen, S.T., Ruoff, R.S., 2006. Graphene-based composite materials. Nature 442, 282–286.

Steim, R., Ameri, T., Schilinsky, P., Waldauf, C., Dennler, G., Scharber, M., Brabec, C.J., 2011. Organic photovoltaics for low light applications. Solar Energy Mater. Solar Cells 95, 3256–3261.

Steirer, K.X., Reese, M.O., Rupert, B.L., Kopidakis, N., Olson, D.C., Collins, R.T., Ginley, D.S., 2009. Ultrasonic spray deposition for production of organic solar cells. Solar Energy Mater. Solar Cells 93, 447–453.

Strong, A.B., 2006. Plastics: Materials and Processing. Pearson Prentice Hall.

Susanna, G., Salamandra, L., Brown, T.M., Di Carlo, A., Brunetti, F., Reale, A., 2011. Airbrush spray-coating of polymer bulk-heterojunction solar cells. Solar Energy Mater. Solar Cells 95, 1775–1778.

Syed, A.A., Dinesan, M.K., 1991. Review: polyaniline—a novel polymeric material. Talanta 38, 815–837.

Szentes, A., Varga, C., Horvath, G., Bartha, L., Konya, Z., Haspel, H., Szel, J., Kukovecz, A., 2012. Electrical resistivity and thermal properties of compatibilized multi-walled carbon nanotube/polypropylene composites. Express Polym. Lett. 6, 494–502.

Tekin, E., Smith, P.J., Schubert, U.S., 2008. Inkjet printing as a deposition and patterning tool for polymers and inorganic particles. Soft Matter 4, 703–713.

Thomas, J.P., Qidwai, M.A., 2004. Mechanical design and performance of composite multifunctional materials. Acta Mater. 52, 2155–2164.

Tiusanen, J., Vlasveld, D., Vuorinen, J., 2012. Review on the effects of injection molding parameters on the electrical resistivity of carbon nanotube filled polymer parts. Compos. Sci. Technol. 72, 1741–1752.

Tjong, S.C., Liang, G.D., Bao, S.P., 2007. Electrical behavior of polypropylene/multiwalled carbon nanotube nanocomposites with low percolation threshold. Scr. Mater. 57, 461–464.

Townsend, P.D., Bradley, D.D.C., Horton, M.E., Pereira, C.M., Friend, R.H., Billingham, N.C., Calvert, P.D., Foot, P.J.S., Bott, D.C., Chai, C.K., Walker, N.S., Williams, K.P.J., 1985. Electronic properties of polyacetylene prepared by the durham route. In: Kuzmany, H., Mehring, M., Roth, S. (Eds.), Electronic Properties of Polymers and Related Compounds. Springer, Berlin, Heidelberg.

Verhelst, W.F., Wolthuis, K.G., Voet, A., Ehrburger, P., Donnet, J.B., 1977. The role of morphology and structure of carbon blacks in the electrical conductance of vulcanizates. Rubber Chem. Technol. 50, 735–746.

Yi, X., Shen, L., Pan, Y., 2001. Thermal volume expansion in polymeric PEC composites: a theoretical approach. Compos. Sci. Technol. 61, 949–956.

Yin, X., Kumar, S., 2006. Flow visualization of the liquid emptying process in scaled-up gravure grooves and cells. Chem. Eng. Sci. 61, 1146–1156.

Yin, Z., Huang, Y., Bu, N., Wang, X., Xiong, Y., 2010. Inkjet printing for flexible electronics: materials, processes and equipments. Chin. Sci. Bull. 55, 3383–3407.

Yoshioka, Y., Jabbour, G.E., 2006. Desktop inkjet printer as a tool to print conducting polymers. Synth. Metals 156, 779–783.

Zhou, J., Lubineau, G., 2013. Improving electrical conductivity in polycarbonate nanocomposites using highly conductive PEDOT/PSS coated MWCNTs. ACS Appl. Mater. Interfaces 5, 6189–6200.

3 Composites: Manufacture and Application

Bethany Middleton

3.1 Introduction

Composites consist of two or more distinct phases, and although they make up a single material, the components remain individually recognizable and separable. The constituents that form the composite material must be present in reasonable proportions, say greater than 5%, and be dispersed uniformly among each other. By combining two (or more) different materials, the resulting composite benefits from the properties of the two constituent phases, and has properties not achievable by individual constituents (Thomas et al., 2012).

Most composite materials generally comprise two phases, a continuous matrix combined with a dispersed reinforcement. There is a wide range of matrix and reinforcement materials resulting in a vast number of composite combinations, each with differing properties and benefits. Although metal and ceramic matrix composites exist, this book focuses on polymer matrix composites (PMCs).

PMCs are often used as lightweight metal replacements and have the advantages of being lightweight, stiff, and strong, and generally have lower manufacturing costs when compared to metals. Considering the huge range of potential polymer matrix materials that can be combined with a number of different reinforcement types, which themselves can be arranged in various different architectures, the range of PMCs becomes apparent.

Before multifunctional aspects can be explored, it is necessary to introduce the properties of composites and their manufacturing methods. Therefore, this chapter begins with the properties of the constituent materials.

3.2 Materials

Most PMCs have two constituent phases; the continuous (primary) phase, known as the matrix, is usually the more ductile of the constituent materials and holds the dispersed phase. The dispersed (secondary) phase

Design and Manufacture of Plastic Components for Multifunctionality
http://dx.doi.org/10.1016/B978-0-323-34061-8.00003-X

is discontinuous in nature, separated, and set within the matrix. It is usually the stronger of the constituent materials and is therefore also known as the reinforcement (Thomas et al., 2012).

3.2.1 Matrix

It may come as no surprise that polymer matrix composites have polymers as the matrix material. However, polymers are a wide ranging group of materials with many different types and chemical families.

Although the matrix is the physically weaker phase of the composite material, it has a number of important functions. It holds the reinforcement in place, maintaining the structure of the part; it protects the reinforcement from abrasion, impact, and environmental conditions; and it transfers the load to the reinforcement.

Polymers are compounds that comprise long chain molecules made up of repeating molecular units ($>10^5$) bonded together covalently. In a polymer, the repeating units can be the same or different, and can the structured linearly or in various branched architectures. They are chemically polymerized from small molecules derived from resources such as natural gas, crude oil, plant oils, and starches.

Polymers fall into two distinct categories – thermoplastic and thermoset. The difference between them all depends on a curing or crosslinking reaction that either does or does not occur during the molding process. (Also see Chapter 4 for more information on this.)

Once thermoplastic polymers have been synthesized (by polymerization) they do not undergo any further chemical reaction. They are supplied in a fully polymerized solid state to the manufacturer. When they are processed or molded they are melted, formed, and cooled to solidify the shape. Since this is only a physical change, the melting, forming, and cooling cycle can be repeated, meaning thermoplastics can the reprocessed and recycled.

The molecular chains in thermoplastics can be completely amorphous (randomly arranged) or they can have some ordered regions that can improve the materials physical properties. The ordered or crystalline regions are those that give semicrystalline thermoplastic polymers their name.

Thermoplastics are generally tough with good impact properties and are recyclable; however, they can be limited in high-temperature applications.

The two most common thermoplastics that are used as the matrix phase of SMCs are polyamides (PA) and polypropylene (PP).

Thermoset polymers are supplied to the manufacturer in a partially polymerized state and can be in a liquid or dough-like form. When they are processed they are molded into the required shape and a curing or crosslinking reaction initiates. This is a chemical reaction that covalently bonds together the shorter chains of the partially polymerized material into a strong interlinked molecular network. The crosslinking reaction can be initiated by temperature, UV light, or a catalyst.

Since a chemical reaction has occurred during the molding process, once cured thermoset materials cannot be easily reprocessed or recycled.

Generally, thermosets are stronger and stiffer than thermoplastics; however, they can be more expensive and take longer to process.

Some thermoset polymers that are commonly used in PMC are epoxies, phenolics, and polyesters.

Both types of polymers are used as matrices in composites and therefore unsurprisingly the resulting PMCs are called either thermoplastic composites or thermoset composites. Each has their own benefits and drawbacks, which will be discussed in a later section.

In the following section, the different types and structures of reinforcement will be described.

3.2.2 Reinforcement

In composite materials, the reinforcement phase can be in any form, such as particles, fibers, or sheets; however, in PMCs the predominant state of reinforcement material is fibrous and as such this is the only type that will be described in detail in this chapter.

The fibrous nature of the reinforcement in PMCs gives rise to other commonly used names:

- Fiber reinforced polymers or plastics (FRPs), and
- Fiber reinforced thermoplastics (FRTPs)

An additional letter in front of the acronym can indicate the type of fiber in the composite. For example, CFRP stands for carbon fiber reinforced polymer.

Even though we are not describing particulate or sheet reinforcement here, this does not mean that the range of reinforcements is small. Fiber reinforcements can vary by length, arrangement, and, of course, material type.

First, fibers can be considered short, long, or continuous. Short fibers are classified as those with a length-to-diameter ratio of less than 100 and are generally less than 2 mm. In short fiber PMCs, fibers are usually randomly dispersed within the polymer matrix. Due to the short length of the fibers, the improved properties imparted to the material from the fibers are limited, but this can be improved by aligning the fibers in the loading direction. This is illustrated in Figure 3.1.

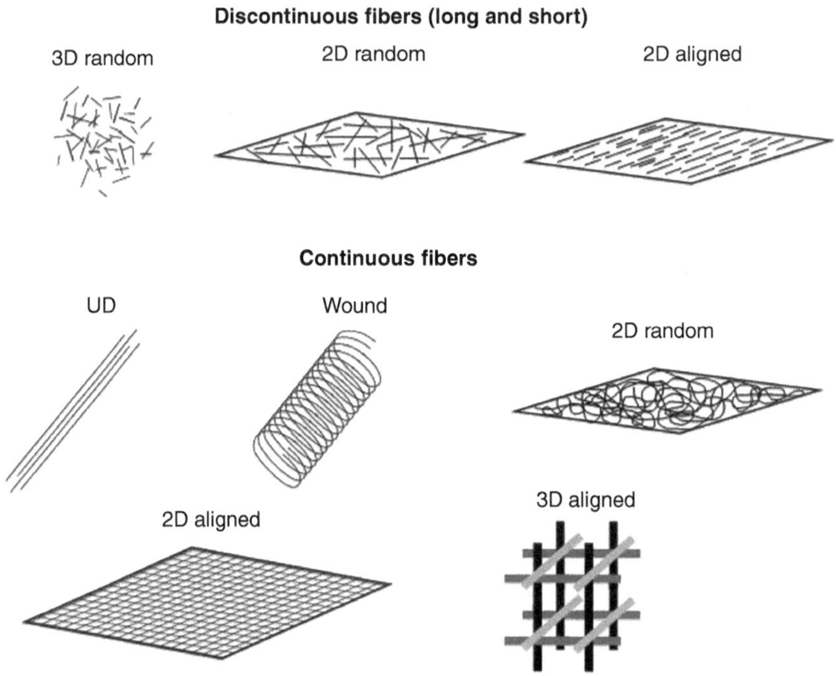

Figure 3.1 Fiber architecture diagram.

Long fibers are obviously longer than short fibers but are still discontinuous in nature. Long fibers can also be randomly dispersed within the polymer matrix, and they can be part of a chopped strand mat that has been processed into a laminate material.

Continuous fibers are those whose length runs the entire length of the manufactured part. Within the composite part the fibers can be unidirectional or wound/spiral. Continuous fibers can also be part of a laminate structure constructed from layers of 2D mats or fabrics. Less common continuous fiber reinforced structures are those formed of 3D woven structures.

Laminate structures are a common form of PMC and as such they are described in more detail here. Some examples of single layer mats and fabrics that can form a composite laminate are UD tapes, biaxial, woven, noncrimp fabric, and continuous strand mat. Multiple layers or plies of these materials are stacked into a laminate structure with plies arranged with the different fiber directions related to the axial or y-direction; 0°, 90°, +45°, and −45° are most common. This is illustrated in Figure 3.2. Any fiber direction angle can be used; however, this can result in a very complex fiber architecture and lay-up.

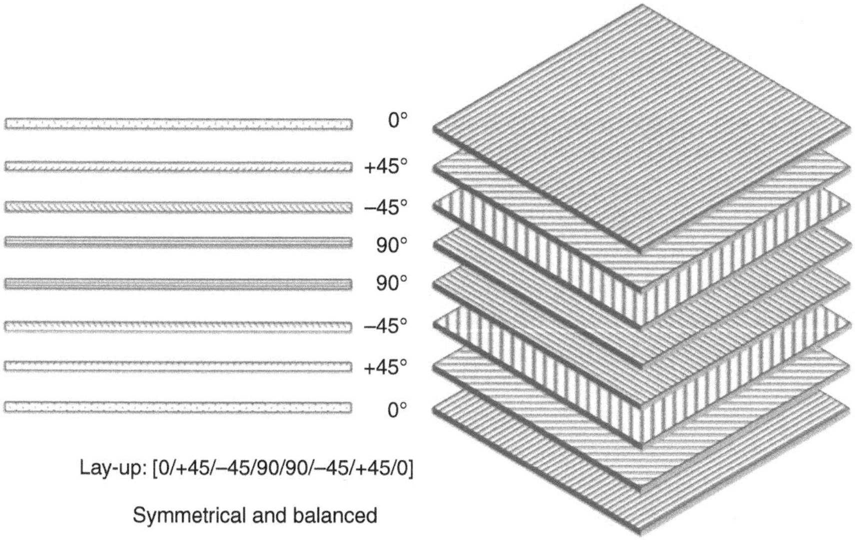

Figure 3.2 Laminate diagram.

When constructing a laminate it is important the lay-up is both symmetrical and balanced. By symmetrical, this means that the ply fiber direction pattern is mirrored about the mid plane of the laminate. This prevents the laminate from warping and bending. A balanced laminate comprises an equal number of $+\Theta$ and $-\Theta$ plies and this reduces shear coupling.

When discussing laminates, the x, y, z coordinate system is used to describe the different directions of the composite material. The axial or 0° direction, along the composite laminate, is described as the "y" direction, the transverse (90°) direction is described as "x," and the "z" direction is through the thickness of the laminate.

One limitation of a laminate composite is its limited physical properties in the z-direction that is the interlaminar shear direction causing delamination of the layers and poor impact resistance. A technique used to overcome this challenge is z-direction reinforcement that has led to the development of three-dimensional (3D) fiber reinforcements for composite structures to overcome this. These include 3D woven composites, stitched composites, and z-pinned composites.

Now, the fiber architecture has been discussed we will move onto the fiber material. Some common types of fiber material used in PMCs are carbon, glass, aramid, and to a lesser degree basalt. Natural fibers and self-reinforcing fibers are also frequently used reinforcements in PMCs. A short description of each follows.

3.2.2.1 Glass Fibers

Glass fibers are the most common type of reinforcement used in PMCs; they generally have good specific strength and stiffness properties and are inexpensive. Glass fiber is made by melting the constituent materials together and drawing the melt into a fiber. They have a diameter of 5–25 μm and are coated with a "sizing" to improve the adhesion with the matrix material. The predominant constituent of glass fiber is silica (silicon dioxide, Si_2), or sand; however, there are different types of glass fiber, each with slightly different compositions and properties. The three most common types of glass fiber are E-glass, C-glass, and S-glass; there are others available but they are usually associated with a particular manufacturer.

E-glass (electrical) is the most commonly used fiber in PMCs. It has good strength and stiffness but poor impact properties. The "E" refers to the electrical resistance of the fibers. C-glass (chemical or corrosion) fibers have good chemical resistance and are often used as a veil or tissue outer layer on pipes or tanks rather than as a reinforcement. S-glass (stiffness) is a glass fiber with improved mechanical and thermal properties. T-glass and R-glass are other types of improved property glass fibers, which along with S-glass cost around 10 times as much as E-glass.

Due to their low cost, glass fibers are typically used as reinforcement in large structures such as wind turbines blades and boat hull.

A quartz glass fiber (10 μm fiber diameter) can be manufactured with a high level of silica; it has a high tensile strength but is expensive.

3.2.2.2 Carbon Fibers

When taking into account the widely used reinforcements, carbon fiber is the most expensive but it has the best specific stiffness properties. It is made by carbonizing a filament/fiber of a precursor material in controlled conditions. The most common precursor used for the manufacture of commercially available carbon fibers is polyacrylonitrile (PAN). The manufactured carbon fibers are then coated with a sizing to protect the fibers and improve adhesion with the matrix. The carbon fibers produced using this method have diameters ranging from 5 to 7 μm.

Carbon fiber is available in different grades, categorized by the modulus of the fiber, increasing in cost as the modulus increases. The standard modulus fibers are referred to as high strength (HS); then there are intermediate modulus (IM), high modulus (HM), and ultrahigh modulus (UHM). Although carbon fiber has very good tensile strength and modulus, they have low impact strength and the highest modulus fibers can be brittle.

Due to its high cost, carbon fibers are typically used in aerospace and high-end automotive applications.

3.2.2.3 Aramid Fibers

Aramid fibers are organic polymer fibers synthesized by spinning the product of the constituent chemical solutions, and like the other fibers that have been described so far, it comes in different grades with a fiber diameter of around 10 μm. Since it is a polymeric fiber it has low density compared with other comparable fibers and therefore high specific properties. The name comes from "aromatic polyamides" and despite being a polymer the fiber has very good thermal resistance and mechanical properties. They have particularly good tensile strength and impact properties that is why they are commonly used in ballistic applications such as body armor and climbing rope. A common trade name of aramid fiber is Kevlar by Dupont.

3.2.2.4 Basalt Fibers

Similarly to glass fibers, basalt fibers are made by drawing a fiber from a molten material. However, in this case, the melt is 100% basalt with no further constituent ingredients or additives. Therefore, they are considered considerably more ecofriendly than carbon fibers. They have a long history of engineering application use since the 1920s. Basalt fibers are typically 9–13 μm and have a higher strength and stiffness than glass fibers and are not as expensive as carbon. They are mainly used in civil applications needing superior properties to glass but despite being not as widely used as either glass or carbon appear to offer considerable further multifunctional potential uses in conjunction with other materials.

3.2.2.5 Boron Fibers

Boron fibers are actually a coated metal filament with extremely high strength, stiffness, and thermal resistance. However, boron fibers are very expensive due to their manufacturing process and substrate, and as such they are only used in PMCs in specialized applications. Boron is deposited onto a tungsten filament producing a >100 μm thick fiber with a boron outer surface. The high price limits their use to high-temperature aerospace applications.

3.2.2.6 Natural Fibers

Natural fibers are low-cost, low-density fibers generally derived from plant materials. Some examples of natural fibers used in PMCs are jute, sisal, flax, and hemp. They can have fiber diameters of up to 500 μm and within each fiber type there is a wide variety in fiber properties based on differences in growing conditions, harvesting seasons, and location of the fiber in the plant. Natural fibers composites have good impact properties and are often reported to have high specific stiffness; however, compared to materials such as GFRPs, they generally have low absolute strength and stiffness. Natural fiber composites are used in nonstructural applications such as decking, lightweight pallets, and automotive trim and interior panels.

3.2.2.7 *Self-Reinforced Polymers*

Self-reinforced polymers or plastics (SRPs) are thermoplastic composites made using the same material for the matrix and the reinforcement. The reinforcement is a fiber consisting of highly orientated polymer chains that have higher mechanical properties and melting point than the standard polymer. SRPs have excellent impact properties and are potentially 100% recyclable. Polypropylene (PP) is the most common form of SRP with a number of commercial products available. Some examples of SRP applications are hard suitcases and automotive panels.

A comparison of the mechanical properties of some typical fibers used in PMCs is given in Table 3.1.

3.2.3 Composite Materials

Although we have described the different architectures of reinforcing fibers in PMCs, this does not fully explain the different formats that composites and their constituent materials are supplied in.

Short fibers are compounded (see Chapter 1) with the matrix material and supplied pelletized ready for processing using techniques such as injection or compression molding. Long fibers can also be supplied in

Table 3.1 Comparison of Reinforcement Properties Table

Fiber	Tensile Strength (MPa)	Modulus (GPa)	Density (10^3 kg/m^3)	Elongation (%)
E-glass	2400–3500	70–76	2.5	1.8–4.8
S-glass	3450–4570	86	2.5	2.8
HM carbon	2400–3500	325–440	1.8–1.95	0.6
HS carbon	3400–3500	160–270	1.75–1.8	1.1
Aramid	3000–3400	60–120	1.45	3.3–3.7
Polyolefin	2000	1–1.5	0.97	3.5
Flax	350–2000	27–100	1.4–1.5	1.3–3.5
Jute	400–800	10–75	1.3–1.5	1.5–1.8

a pellet form; to maintain the longest possible fiber length (usually a maximum of 10–12 mm) these are manufactured using pultrusion (which is covered in the processing section later), rather than compounding in an extruder. These can also be injection molded, but care is needed when processing to reduce the amount of fiber breakage.

Long fibers are also supplied within the matrix material in the form of bulk and sheet molding compound (BMC and SMC). Both BMC (completely random fiber arrangement) and SMC (random fiber orientation in two dimensions) can be compression molded. Chopped strand mat (CSM) is another form in which long fibers are supplied. This is a two-dimensional sheet of randomly arranged chopped strands held together with a binder, ready to be combined with the matrix during processing. These long fiber products are generally manufactured using glass fibers. Glass mat reinforced thermoplastic (GMT) is a thermoplastic equivalent to the thermosetting CSM. Although the fibers in SMC, CSM, and GMT are described as long, they are generally longer than the "long" fibers processed using injection molding. The lengths of fibers in these materials are typically 25–50 mm.

For continuous fiber composites, there is a range of widely available commercial materials.

Some processes such as filament winding and pultrusion can manufacture composites straight from fiber, so the simplest material supplied are groups of filaments/fibers on a reel. Bundles of glass fibers are known as rovings and carbon fibers are known as tows. Bundles of glass fibers can also be manufactured into and supplied as yarns; the difference between a roving and a yarn is that the fibers in a yarn are twisted together. Rovings and yarns come in different "tex," which is an SI unit where 1 tex =1 g/km of a glass bundle. Tows are described by their size using a "k" number; this describes the number of fibers in the bundle, for example a 6 k tow has 6000 fibers in the bundle.

Unidirectional (UD) fabrics are supplied on a roll; these are constructed to have a majority of the fibers running in the direction of the roll with a small amount of material running at 90° to the roll direction in order to hold the UD fibers together. There are also tapes available that combine a thermoplastic matrix with unidirectional fibers and are supplied on a roll; these are generally called UD tapes.

Bundles of continuous fibers can be supplied as stitched and woven fabrics. Glass fiber fabrics can be constructed using both yarns and rovings. The most common fabrics are constructed from bundles of fibers in the 0° and 90° directions; these are known as 0/90 or biaxial fabrics. These come in many tow/roving sizes and woven fabrics can also be manufactured using different weave patterns.

Stitched fabrics are often called "noncrimp fabrics" since by stitching the fibers together rather than weaving them it removes the "crimp" caused by interlacing the fibers during weaving.

Multiaxial fabrics (noncrimp) are defined by EN 13473 as

> A textile structure constructed out of one or more laid parallel non-crimped not-woven thread plies with the possibility of different orientations, different thread densities of single thread plies and possible integration of fiber fleeces, foams or other materials, fixed by loop systems or chemical binding systems. Threads can be orientated parallel or alternating crosswise. These products can be made on machines with insertion devices (parallel weft or cross-weft) and warp knitting machines of chemical binding systems.

The advantages of such systems are utilized in aerospace with carbon fiber systems, in which the noncrimps can be woven or stitched together to enhance unidirectional layers.

Although most fabrics are constructed from a single type of reinforcement, there are hybrid fabrics manufactured with two different reinforcing fibers. Glass, carbon, and aramid fibers are the types most commonly combined in hybrid fabrics and again there is potential for further hybrid, multifunction to be introduced utilizing methods such as these. Figure 3.3 illustrates the concept.

Once assembled, these fabrics can be supplied "dry" and then be processed with separately supplied resins by RTM or resin infusion. Resins can be supplied as one-part or two-part systems. Fabrics can also be sup-

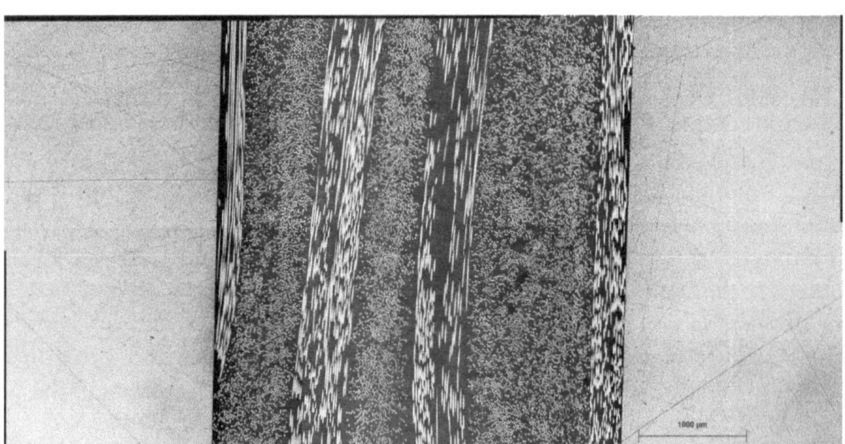

Figure 3.3 Laminate structure micrograph.

plied, readily combined with the resin material. Various examples are shown in Figures 3.4–3.7.

A commonly used commercially available composite material is "pre-preg" which is short for "*pre*-im*preg*nated fabric." This is a fabric that has been pre-impregnated with the resin that can be processed by vacuum bagging or autoclaving and requires no addition of any other material (Figure 3.8). Since the resin in a pre-preg is usually a thermoset, they have a limited shelf life and are required to be stored in a freezer to slow down cure. Thermoplastic pre-pregs are also available where a

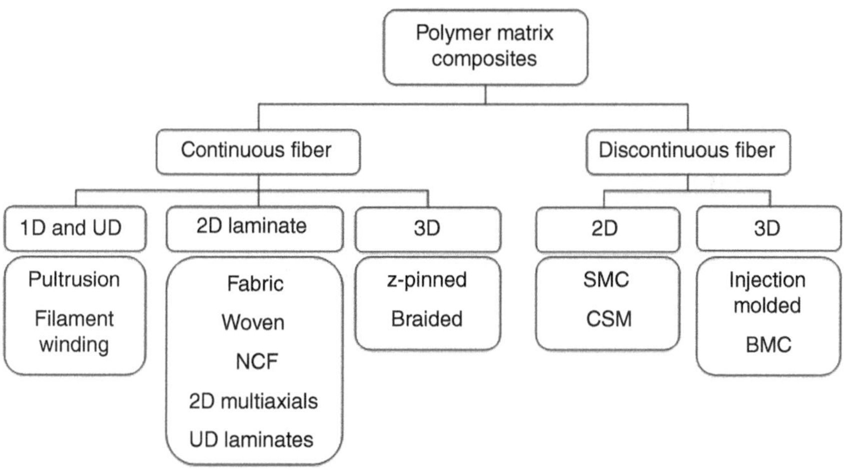

Figure 3.4 Classification of PMCs.

Figure 3.5 Thermoplastic UD tapes.

Figure 3.6 Dry fabrics.

sandwich of TP film–reinforcement fabric–TP film is calendared together (Balasubramanian, 2014).

In order to know the full specification of a pre-preg, the fiber type, the fiber bundle size, the fabric type, the resin type, and fiber:resin ratios, pre-pregs are typically an expensive composite processing material since it has already gone through a number of manufacturing stages to produce it. Pre-preg can be fully or partially pre-impregnated and can be produced by a number of different manufacturing routes such as impregnation with melt, powder or solvent systems, and the commingling technique. An in-depth discussion of these techniques is beyond the scope of this book and the reader is referred to the subjects of textile manufacturing for further information on this area. While it is at these points that the potential to add other materials (such as simple fillers or more complex CNT materials)

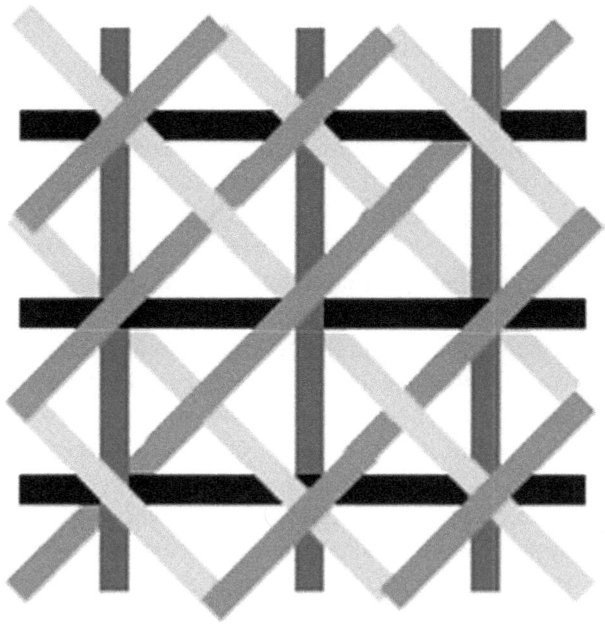

Figure 3.7 Multiaxial fabric photo/diagram.

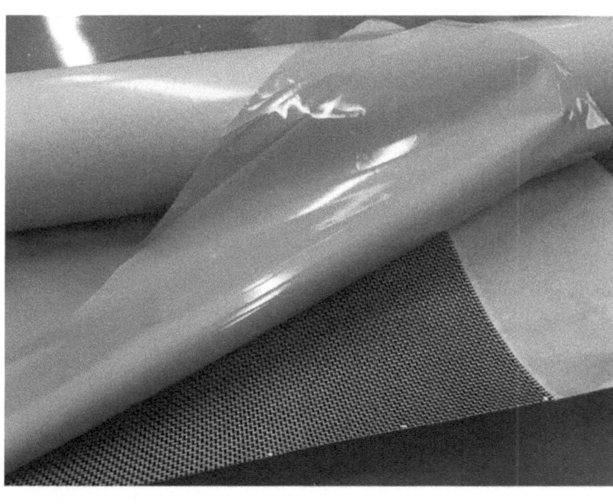

Figure 3.8 Prepreg.

exists, these specialized processes mean it is necessary to talk to the prepreg manufacturers themselves.

For this reason, it is the authors' opinion that other in-situ so called "reactive" processes such as resin injection molding (RIM) and resin transfer molding (RTM) offer more potential for "play" and more customized processes that will be covered in the processing section.

Nonwoven matrix material veils and tissues are also available and can be used at the surface to improve the surface quality of the manufactured composite or can be part of a complete system, multilayered fabric. In these fabrics, the matrix is incorporated by using veils of matrix material that are positioned in between reinforcement layers that can be in a woven or NCF form. Thermoplastic composites can also be manufactured as an intermingled fabric where thermoplastic matrix fibers and reinforcement fibers are present alongside each other within each bundle of the fabric.

To obtain even greater weight savings while maintaining good mechanical properties, composite materials are often used to manufacture sandwich structures. Typical core materials for sandwich structures are foams and honeycombs. These can be applied to a number of the processes that will now be discussed.

3.3 Processing

There are a large number of PMC processing methods that range from completely manual and skill dependent to entirely automated, with various degrees in between. Each processing method has its own advantages and challenges, meaning that no one method is necessarily better than the other, but its use is entirely dependent on the part being manufactured. All of the methods are used today and a number of different varieties of many of the processing types have been developed, often associated with (and proprietary to) particular manufacturers. In this chapter, an overview of each basic processing method is given.

3.3.1 Hand Lay-Up

Hand (or wet) lay-up is the most basic and lowest cost thermoset composite processing method when considering the equipment used; however, the quality of parts produced is highly dependent upon the skill of the processor, and therefore this is a highly skilled process. The process is shown in Figure 3.9. The reinforcement used is usually a dry fabric, which can be woven or stitched. Operators apply the reinforcement into

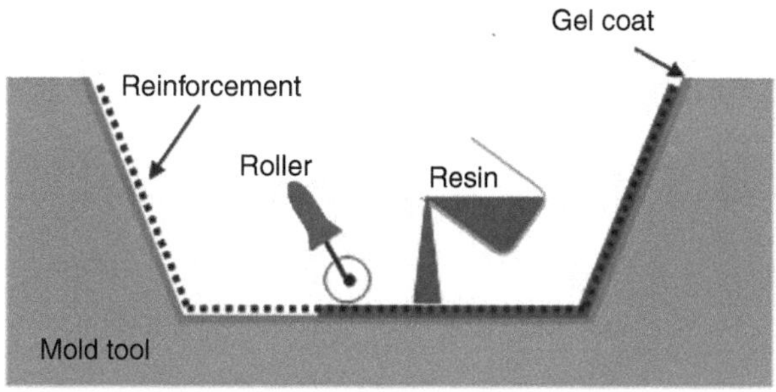

Figure 3.9 Hand lay-up diagram.

a single-sided mold (usually female), carefully draping and manipulating the fabric over the contours of the tool cavity. Resin is applied by hand by either rollers or brushes. The manual application of the matrix means a low-viscosity resin is favored for hand lay-up and wet-out of heavy weight fabric or high fiber volume fraction composites is difficult to achieve. Nip-rollers impregnator systems can be used to force resin into fabrics improving wet-out.

Parts cure under atmospheric conditions and as such the tooling can be low-cost and large, meaning that the size of the part to be manufactured is not limited by equipment. Tooling is often made from composites but can be made from aluminum or steel. Support frames are often needed to support the tool in place. Due to the long room temperature curing regime and the manual lay-up, the overall cycle time is long for this processing method.

Hand lay-up is commonly used by hobbyists for making large structural parts such as boat hulls and wind turbine blades. Sandwich structures can also be manufactured using hand lay-up by introducing honeycomb or other structures into the build.

3.3.2 Spray Lay-Up

Similarly to hand lay-up an operator applies the composite material into single-sided tooling and the composite cures in atmospheric conditions, so it also has the advantage of low-cost tooling, can produce large parts but has a long cycle time. This method differs to hand lay-up by the processor using a hand-held gun to spray on a combination of thermoset resin and chopped fiber at the same time. The gun has a feed of glass roving and resin. It chops the fibers, combines the fibers and resin together, and uses compressed air to deposit the composite as a spray. This results in a high fiber deposition rate. This process is illustrated in Figure 3.10.

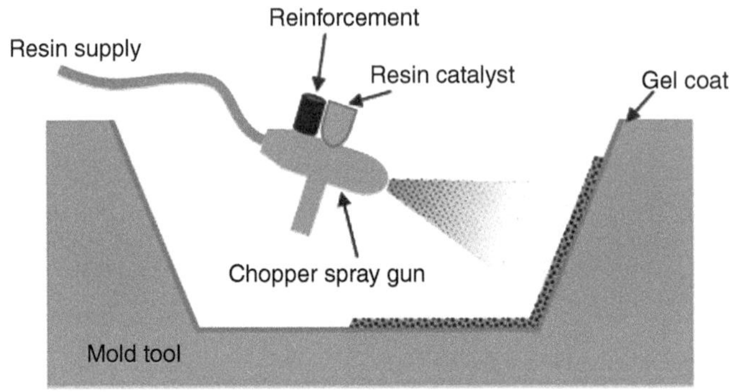

Figure 3.10 Spray lay-up diagram.

The material applied using this method is formed of chopped fibers and is resin rich, so the resulting composite has limited mechanical properties and weight saving. Caravan bodies and vehicle aerodynamic panels are examples of the low-loading structures made using spray lay-up.

3.3.3 Vacuum Bagging

Vacuum bagging (see Figures 3.11 and 3.12) or vac-bagging is a curing method used to produce composites with improved porosity and higher fiber content. A composite that has been constructed using the hand/wet lay-up method is sealed to the tool with a plastic film/bag. The air inside the bag is evacuated using a vacuum pump; therefore, the material inside is subjected to one atmosphere of pressure, resulting in an improved property composite. The vacuum bag system can also be placed inside an oven so that the curing can take place at elevated temperature speeding up the reaction. This also means that pre-preg and thermoplastic composites can be processed using vac-bagging.

Although composite produced using vac-bagging has improved mechanical properties and in a faster cycle time the process is more expensive than hand lay-up due to the large amounts of consumable bagging materials such as bag, release film, breather fabric, and sealant tape. Part size can be limited and tooling can also be more expensive if oven curing is required.

A wide variety of components can be produced including glass fiber and carbon fiber products and sandwich structures can also be made using vac-bagging.

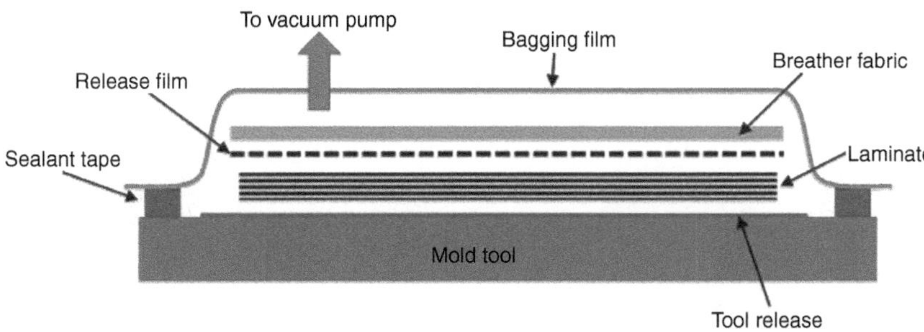

Figure 3.11 VAC bag diagram.

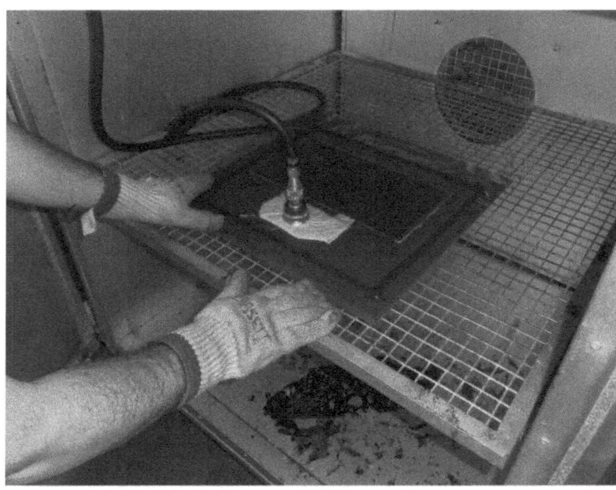

Figure 3.12 Vacuum bagging.

3.3.4 Resin Infusion

In order to manufacture a composite part using resin infusion, a dry stack of fabrics is laid up on single-sided tooling and bagged in a similar way to vac-bagging. Two ports are included in the bag, one from the thermoset resin supply and one to a vacuum pump. The vacuum draws the resin across and through the fabric, resulting in a high fiber content composite; however, to get good impregnation a low-viscosity resin is required for this process. The curing generally takes place at room temperature and as such large tooling (and therefore parts) can be accommodated. For large parts, however, the process can be complex and it can be difficult to produce consistent parts with no dry areas.

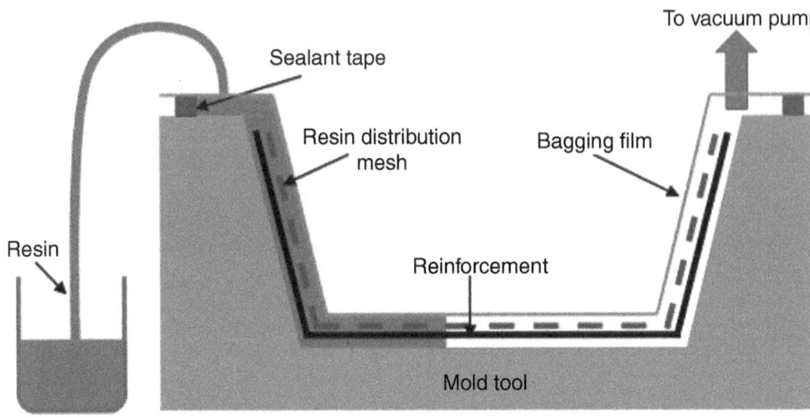

Figure 3.13 Resin infusion diagram.

Yacht hulls, vehicle panels are parts typically made using the resin infusion process that is illustrated in Figure 3.13.

3.3.5 Resin Transfer Molding (RTM)

In RTM, a dry fabric lay-up is placed between a set of matched tooling that is held together using clamps or a press. Depending on the shape of the molded part, the dry fabric may need to be preformed. A thermoset resin is then injected under pressure (typically between 2 and 20 bar) and flows through and impregnates the fabric. The flow of resin can be slow, as such a number of injection points may be used and stitched fabrics facilitate a faster resin flow than wovens. The process is dependent on the pressure gradient in the tool, the viscosity of the resin, and the architecture and nature of the fabric, that is, its inherent permeability. The composite is allowed to cure once the fabric has completely wet out; this can happen at room or elevated temperatures and therefore this time also varies considerably from minutes to hours in some cases.

Resin infusion and RTM both produce composite parts with a low void content, high fiber content, and high mechanical properties. For RTM, since matched tooling is used, both sides of the part have a molded surface/finish and complex 3D shapes can be produced; however, the tooling is expensive and limits the size of the parts being manufactured. The process is shown in Figure 3.14.

RTM can be used to make complex composite automotive parts.

To improve cycle times and part quality and properties, number of variations of RTM have been developed. These include vacuum-assisted resin transfer molding (VARTM) and high-pressure resin transfer molding

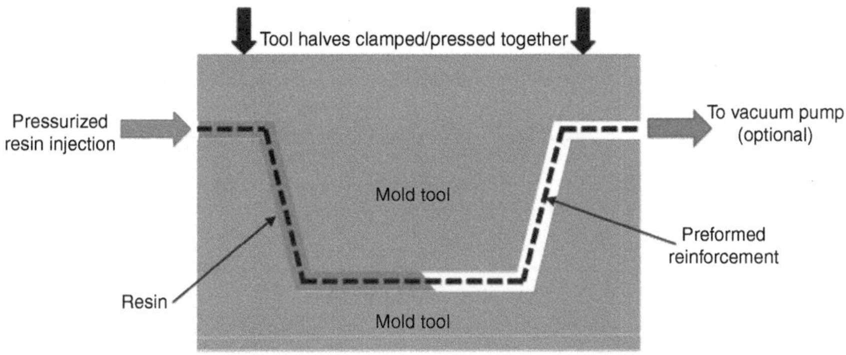

Figure 3.14 RTM diagram.

Figure 3.15 Combined compression and HP-RTM machine.

(HP-RTM). Although these processes have advantages, by adding complexity to them they are more expensive. A hybrid RTM/compression molding machine is shown in Figure 3.15.

3.3.6 Compression Molding

Compression molding is typically used for the manufacture of a small-to-medium randomly aligned fiber reinforced thermoset composite. Compression molded composite components are produced by placing a measured charge of preheated material into the cavity of a heated matching tool set. A hydraulic press closes the mold and the material flows throughout the cavity. Bulk and sheet molding compounds (BMCs and SMCs) are

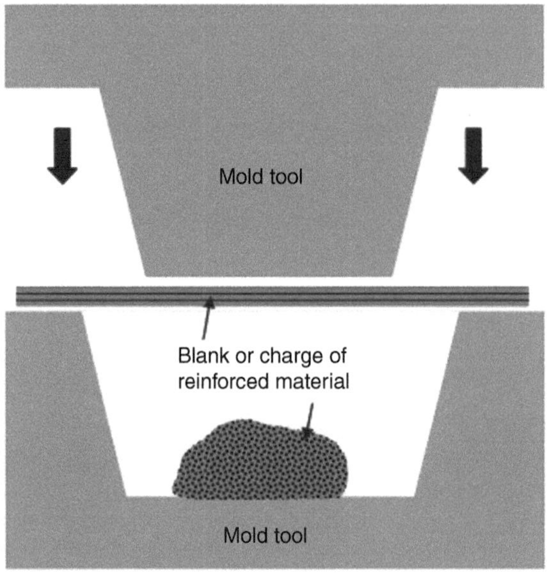

Figure 3.16 Compression/forming diagram.

commonly processed using compression molding, and subsequently cure while the heated mold tool is clamped shut. Due to the use of noncontinuous fiber reinforcement, compression molded BMC components typically have low-mechanical properties; however, SMC moldings have improved properties. Compression molded SMC components can be used to replace metal stampings in a number of different automotive applications with the benefit of being lighter, resistant to corrosion, and requiring cheaper tooling (Bishop et al., 2009). The process is shown in Figure 3.16.

Composite products made using compression molding are less complex in design than injection molding but for multifunctionality are interesting as the tool platen orientation is in a different orientation to injection molding; they are not subject to gravity and therefore it is easier to place inserts into the mold. Over-molding of compression molding components or vice versa over-molding of injection molding components by compression molding are useful variants. Over-molding is a technique used widely in automotive applications.

3.3.7 Stamp Forming

Similarly to compression molding, a matched set of tooling on a hydraulic press is used to stamp form composite material. Stamp forming refers to the rapid forming of laminate composite in a heated press. Pre-preg, thermoplastic tapes, intermingled fabrics, and other fabrics can all be stamp formed using different thermal regimes. The process is very fast (seconds) and has the benefit of producing parts with high mechanical properties due to the continuous nature of the fibers within these parts. The size of the parts is limited by the size of the press. Typical PMC parts made using stamp forming are high-performance plastics such as PEEK, PEI, and PPS for high-end applications such as military and aerospace. The materials are preheated in an oven, transferred to the press, and then rapidly stamped.

3.3.8 Automated Cutting And AFP

One of the driving forces in composite processing is to move away from very manual intensive manufacture and toward more automated processing (Figure 3.17). Precutting of preforms and automated layup is, therefore, a very important part of the composite industry; however, this area has very little relevance to the main focus of this book. The reader is, therefore, referred to more specialized composite titles to cover this aspect of composite research or to the Robotics and automation section in Chapter 4.

3.3.9 Autoclave

Autoclaving is one of the most expensive composite-processing methods due to the high cost of the autoclave itself along with the high price of the pre-pregs that are typically processed using this method. However, it

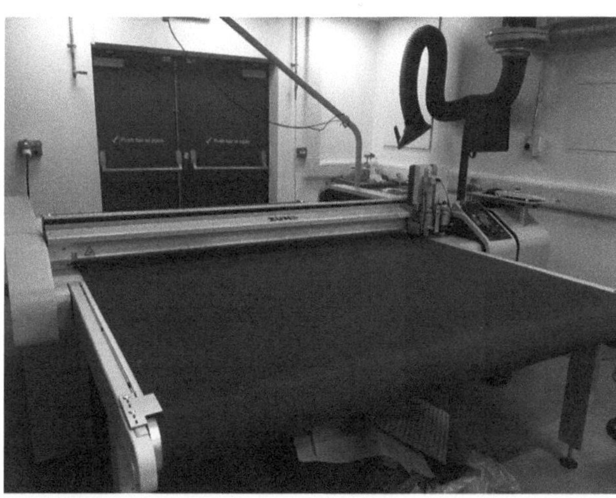

Figure 3.17 Automated cutter.

does produce composite parts with the highest mechanical properties and is therefore used in applications with the highest load requirements.

Pre-pregs are laid up (manually or automated) into laminates in the same way as a vacuum bagged composites; however, consolidation or "debulking" is carried out every few plies. The bag is evacuated and then placed inside the autoclave for curing (Figure 3.18). The autoclave applies elevated temperatures and pressures to produce a high fiber content composite with a low void content. A typical cure regime is 5–6 atmospheres of pressure and 120–180°C. The tooling is expensive since it is required to be used at temperature. Invar is a popular material for tooling due to the low coefficient of thermal expansion.

An autoclave is shown in Figure 3.18. Structural parts made using the autoclave process are aircraft wing structures, Formula 1 racing car components, marine structures, and also including sandwich panels.

3.3.10 Out of Autoclave (OoA)

The high temperatures and pressures that the autoclave process requires result in high processing costs and a capital investment in the purchase of an autoclave that can be many thousands of pounds for large components. Some pre-pregs have now been developed to be cured "out of autoclave" (OoA) or by "vacuum bag only" (VBO). This results in a lower cost production process, lower cost tooling, and it is claimed that it produces autoclave quality parts. This is a material rather than process development, but OoA materials are an area of great commercial interest for their ability to reduce costs and manufacturing speeds.

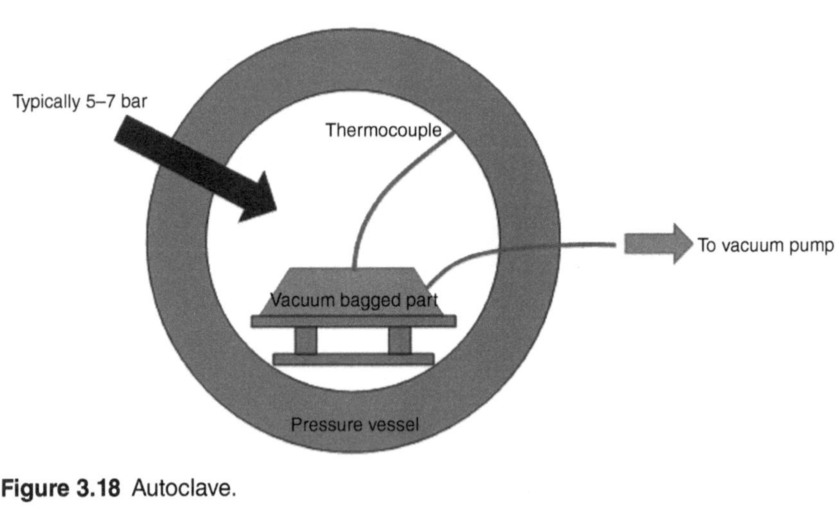

Figure 3.18 Autoclave.

3.3.11 Injection Molding

In this chapter, injection molding will only be mentioned briefly in how it relates specifically to the processing of composites. But since it is such an advanced and versatile process it has a chapter all to itself. For a more detailed description of the injection molding machine and process, refer to Chapter 4.

Injection molding is a fast, highly automated method of producing short fiber reinforced PMC parts with complex geometry. Standard injection molding produces thermoplastic composite parts but thermoset composites can also be manufactured. Short fiber reinforced thermoplastic is melted and homogenized by a rotating screw in a heated barrel. The melt is then injected into a set of matched tooling at high pressure where the matrix cools. The alignment of the fibers depends upon the flow direction of the material; however, it is partly made using injection molding generally described as having random alignment. Since the fibers are short and in a random orientation the mechanical properties are generally low. Due to the complexity of the injection molding process the capital investment is high and the tooling is expensive, because of this injection molding is usually limited to relatively small, high volume parts.

Injection molding is used to make high-volume, nonstructural parts such as automotive, engine covers, floor pans, and body panels. An example machine is shown in Figure 3.19.

A variation in injection molding, called reactive injection molding (RIM), manufactures thermoset parts using the same basic process. The only difference is that a fast curing thermoset resin is mixed with

Figure 3.19 Injection molding.

a catalyst in a mixing head before being injected into the mold. Two further adaptations to this process are reinforced reactive injection molding (RRIM) and structural reactive injection molding (SRIM). In RRIM short fiber reinforcements are incorporated into the injected resin and in SRIM a reinforcing fiber mesh is placed in the mold and injected over with the resin.

3.3.12 Extrusion

Extrusion is another process used to produce fiber reinforced thermoplastic composite parts. Similarly to standard injection molding, short fiber reinforced thermoplastic is melted within a heated barrel. A screw continually rotates, forcing the melt through a die to form a continuous part with a fixed cross-section. The extrudate is cooled in water to solidify and is then either sectioned into lengths or rolled into a coil.

Again since the fibers are short and have limited alignment, the resulting product has limited mechanical properties. The manufacturing process is also restricted to producing parts with a constant cross-section. A common composite product manufactured using extrusion is wood-plastic decking boards; they are also used to create window profiles.

More detail on screw-based processes can be found in Chapter 1.

3.3.13 Pultrusion

Pultrusion is also a process used for making parts with a constant cross-section. It differs from extrusion in the form of the reinforcement and the fact that the material is pulled through the die rather than being pushed through like extrusion. The use of the die closely controls the resin:fiber ratio and pultrusion has the ability to manufacture composite with a high fiber ratio.

Fiber rovings (sometimes along with fabrics) are drawn from the rear of the machine; they are guided into position, passed through a resin bath and then through a die. Since thermoset materials require time for curing, a mold is then clamped around the material. When this material has cured the clamp opens and more material is pulled through. Therefore, the process is not continuous in the same sense as extrusion; however, long lengths of composite can be made rapidly. A diagram of pultrusion is shown in Figure 3.11.

Since the composite made using pultrusion has a high proportion of fibers on the 0°, it has very good mechanical properties in the longitudinal direction. Beams and girders are examples of products made using pultrusion.

Pultrusion is a process that can also be carried out using thermoplastic as the matrix material.

3.3.14 Filament Winding

Similarly to pultrusion, filament winding uses fiber rovings directly from the creel and passes them through a resin bath. These coated fibers are then wound around a mandrel to form a hollow tube or a container. The speed and orientation of fiber deposition can be controlled by altering the rotating speed of the mandrel and the angle of the fiber feed head.

Mandrel and capital investment costs are expensive and the shapes that can be produced using filament winding are limited, but this process manufactures products with high mechanical properties. Since the fibers are aligned around the axis of the final product, the strength and integrity of the part is very high and therefore products made using filament winding can withstand high internal pressures. Storage tanks, pipes, and gas cylinders are all product made using filament winding.

3.4 Composite Properties and Markets

3.4.1 Bulk Properties

The bulk properties of a composite are different to the properties of the individual constituent materials but benefit from the properties of both the matrix and the reinforcement. For example, a PMC can be stiff and strong, due to the reinforcement properties, and tough and impact resistant, from the matrix properties. In a correctly formulated and manufactured composite, load transfer allows the matrix to transfer stress and strain to the reinforcement. As well as the composite properties being due to the constituent material properties, they also depend on the structure and the proportion of each.

The properties of a composite material can be modeled, calculated, and predicted if the properties, orientation, and proportions of each phase are known; however, a detailed micromechanics approach is beyond the scope of this chapter. The use of the global Cartesian coordinate system allows very complex stresses and strains in individual layers of the structure to be described mathematically and therefore the overall predicted properties are derived. For simpler unidirectional composite structures, a less complex rule of mixtures can be applied to predict certain properties and performance, particularly the modulus in the longitudinal direction.

However, the properties of the composite are not just dependent upon the properties of both the matrix and the fiber, the ratio of both phases, and the fiber orientation, the fiber–matrix interface is also key in allowing the composite to transfer stresses effectively. Also, the manufacturing route is vital as any manufacturing defects such as poor consolidation (voids), misalignment of fibers, contamination and/or other stress raisers, fiber damage, etc, all detrimentally affect the overall properties of the composite.

Of the two phases, the reinforcement is the stronger and stiffer material and in fibers the highest mechanical properties are along their length. Therefore, higher properties are imparted to a composite material with longer fiber lengths and in the direction of the fiber alignment. Fibers have high strength because they have less surface defects due to a small diameter (Campbell, 2010). So randomly aligned short fiber composites have the lowest mechanical properties and continuous fiber UD composites have the highest mechanical properties (at 0° to the fiber direction).

The ratio of the fiber to the matrix also affects the overall property of the composite, the higher the percentage of reinforcement, the stiffer and stronger the resulting composite material. However, this relationship does reach a limit – a minimum percentage of matrix is required to bind the reinforcement together, hold the fibers in place, and distribute the load throughout the reinforcement.

Another important factor responsible for the properties of a composite material is the quality of the interface between the reinforcement and the matrix. Poor adhesion between the two phases in a composite material results in premature failure at the interface, causing the composite to have poor mechanical properties. When a composite has a poor fiber–matrix interface, it often fails by delamination and fiber pull-out. To overcome poor fiber–matrix interface, fibers are coated in sizing to improve the adhesion between the two phases.

Randomly aligned short fiber PMCs have comparable properties in all directions, whereas UD composite laminates have excellent properties in the *y*-direction but low properties in the *x*- and *z*-directions. This is because the property of the UD composite in the *y*-direction is predominantly due to the fiber properties and in the *x*- and *y*-direction the composite properties are mainly determined by the matrix properties.

Since the highest mechanical properties are imparted to the composite in the direction of the fiber, a biaxial laminate PMC would have good properties in both the *x*- and *y*-direction; however, these will only be approximately half the value of equivalent UD composite properties since only half of the fibers are orientated in each direction. Although biaxial laminates have good on-axis properties, in the 0°/90° (or *y*- and *x*-directions),

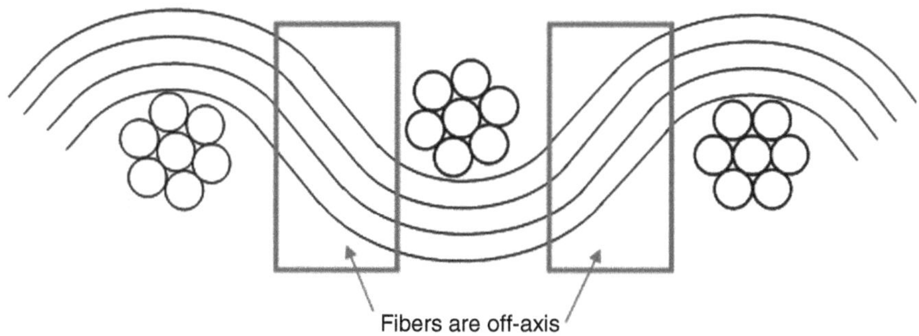

Figure 3.20 Off-axis crimp in woven fabrics.

they do not perform as well off-axis (+/−45° direction). To overcome this, a multiaxial laminate could be used with fibers in the +/−45° direction; however, performance on-axis would be reduced due to the lower proportion of fibers orientated in those directions.

When fibers are woven into a fabric the fibers deviate from the true *y*- or *x*-axis and as such the properties of the composite in those directions are not as high as UD fibers in the same direction. The size of the tow in the woven fabric has an influence on this effect. A material with a large tow size will have a greater reduction in its properties (compared to a small tow size fabric) because there is a greater deviation from the axis as the fiber bundle weaves between the diameter of the large bundles of fibers in the other direction (Figure 3.20).

Since stitched, or noncrimp fabrics, are not woven this reduction in properties caused by the weave pattern or crimping of the fibers does not occur and they generally have higher mechanical properties in the *x*- and *y*-directions than their woven counterparts.

For a similar reason PMCs, or more specifically FRPs, have better tensile properties than compression. If we take UD composites for example, although they are described as UD, there will always be some slight deflections or deviations of some fibers. When in tension, the load pulls those fibers closer to the loading axis, maintaining the strength of the composite by keeping the fibers aligned to the load direction. However, in compression, the effect is opposite to push the fibers out of alignment, thus reducing the properties of the composite in the direction of the load. Also, for this reason, woven fabrics have poorer compression properties than stitched fabrics because the weave pattern causes the fibers to be out of alignment with the load path.

Since the mechanical properties of a composite are dependent on fiber type, quantity, orientation, and interface, the composite material constituents and lay-up can be tailored to have the required properties for particular parts and load paths. For those looking for performance testing, an extensive array of international standards (ISO) exists for evaluating the performance of composites for both mechanical and environmental in service conditions. As these materials often provide such long service life and durability under the harshest conditions, understanding the long-term composite behavior under service loading is of paramount importance.

3.4.2 Composite Market

As well as looking at the science and engineering aspects of PMC materials and applications, it is also important to consider the bigger picture of the worldwide composite markets and the drivers behind the growth in that market. The next section offers a brief overview of the composite market, specifically focussing on the UK.

3.4.2.1 Global and European Composites Market

From 2009 to 2014, by sector, the global composites market was expected to increase annually by 15.6% for aerospace and 13.3% for wind energy. The next big growth areas were pipe/tank manufacture, marine, construction, and transportation all at over 5% annual market increase and followed by a modest growth in consumer goods, electrical, and other sectors (BIS, 2009)

A 2013 Composites Market Report conducted for AVK (the German Federation for Reinforced Plastics) and Carbon Composites e.V. (an association whose members accounted for 40% of CF production in 2012) describes the European GF market and the Global CF market. It states that volume-wise, GF (short, long, and continuous) are the reinforcing fiber used in more than 95% of global FRP composite production. All other types of reinforcement such as carbon, aramid, natural, etc. account for less than a 5% volume of all FRPs. From 2009 to 2013, European GFRP (not including short fiber) production grew from 815 kt to 1020 kt, with around a quarter of that being BMC and SMC. Open processes, such as hand lay-up, are generally in decline, whereas more automated processes show a growth (e.g., RTM grew 5% in 2013). GFR thermoplastics also show a strong growth mainly due to its application in the automotive and transport industry. In 2013, 68% of European FRP production was used in construction and transport applications. In the same year, Germany was the largest producer of FRP with a production of 192 kt (18.8%) and the UK was responsible for more than 13%.

From the same report, it was stated that global demand for CF in 2012 was 43,500 t. This was expected to rise to 68,000 t in 2015 and 130,000 t in 2020. In 2013, PAN-based CF production capacity was dominated by

Europe (24%), USA (23%), and Japan (23%). Of all manufactured CF, 97% is processed into composite materials, and of that 95% go into manufacturing FRPs. As such, the global CFRP market is expected to rise from 65,000 t in 2012 to 208,000 t in 2020 and it is expected that pre-pregs will continue to dominate the market. In 2012, 54% of CF was used in the manufacture of pre-pregs (UD and woven fabrics), 15% was used in winding processes, and 8% used in pultrusion. Short-cut fibers accounted for 18% of CF use and 5% was used in the manufacture of braids and dry fabrics.

Wind energy blades are the single biggest application for CFRPs (23%), followed by the aerospace and defence sector (18%), and the sport and leisure industry (17%). Despite only accounting for 18% of volume used, aerospace and defence account for 40% of sales since these applications require higher specification and quality of CF composites that have a higher value. In 2012, Europe was responsible for 46% of the CF consumption in the aerospace and defence sector, 74% of the wind turbine sector, 56% of the automotive sector but just 4% of the sport and leisure sector. While current growth is driven by aerospace, defence, sport/leisure, and wind turbines, it is expected that automotive has the greatest potential for growth in the future.

3.4.2.2 UK Composite Market

Country-by-country figures will vary; therefore, a focus on just one individual country, the UK, now follows. According to a UK Trade and Investment report published in 2008, the production value of composite components manufactured in the UK in 2010 was estimated to be £453m (40%) for GF composites and £658m (60%) for CF composites. By volume this equates to 87 kt on GF and just 2.5 kt of CF components. This shows that although carbon fiber composites often capture the headlines more than glass fiber composites, a majority of components are manufactured with GF composites. In fact by sector, the production value of the aerospace sector represents 61% of the total UK composites market; however, this only accounts for 1% of the weight of the 2010 UK composite production.

This study identified 39 UK companies as earning >£5m in composites revenue, a further 142 where a significant proportion of their income comes from the composite supply chain and an additional 1300 active companies in the UK sector.

The report predicted that from 2010 to 2020 the UK production demand for GF composites increased from £545m to £1088m and production demand for CF composites will be increased from £707m to £1151m driven by growth in the aerospace and wind energy sectors (UK Trade and Investment, 2010).

Although CFRPs represent a small proportion of the total volume of UK composites market, due to high value of CF, it is the fiber type that represents the majority of the UK composite market monetary value. Due to this, we will look at the UK CF Market in some more detail.

In May 2009, NetComposites (on behalf of the Department for Business Enterprise and Regulatory Reform) compiled a public market profile report on "The UK Carbon Fibre Composites Industry" (Bishop et al., 2009), which states that the UK has a large CF pre-preg sector manufacturing parts for aerospace and motorsport applications. Of the 2.5 kt for CF process in the UK in 2009, roughly 50% goes toward making components in the UK and 50% is exported as high-grade pre-preg fabrics.

For example, in 2008, the CF sector employed over 18,000 people in the UK. The UK manufactured a small amount of CF filament and imported CF in filament form, fabrics, and pre-pregs. The UK imported 1860 t of continuous filament CF; this combined with 200 t of locally produced fiber went on for further processing into finished components and semifinished materials such as fabrics and pre-pregs (UD and fabric). 450 t of continuous filament was processed directly into components using processes such as filament winding and a total of 600 t of fabrics and pre-pregs (of which 440 t are imported) are used to construct composites components; along with 30 t of chopped strand components totals 1180 t of CF component manufactured in the UK. The remaining CF fabrics and pre-pregs (1200 t) were exported for manufacture into components elsewhere around the world.

Of the composite components manufactured in the UK in 2008, 42% were manufactured using pultrusion and filament winding (filament processing) and 46% were laminate lay-ups processed either with or without autoclave. The remaining components were manufactured using compression and injection molding or wet molding or infusion.

The end use of such components was 36% for the aerospace and defence sector, 33% for the wind energy sector, 12% for automotive applications, 7.5% for marine, 7% for industrial (e.g., construction and infrastructure), and 2% and 2.5% for sporting and "other" applications, respectively. This provides a very useful "snapshot" of composite markets.

3.4.2.3 Applications

As can be seen by the sector-by-sector description and processing routes available, polymer matrix composites are used in a wide range of non-, semi-, and structural applications due to the large variety of materials available, ranging from high-cost, high-performance CFRP laminates to low-cost, randomly aligned, resin-rich PMCs. PMCs also have the benefit of a wide range of available processing technologies and relatively low processing costs, resulting in greater scope of shapes that can be manufactured. Another characteristic that all composites have in common is their light weight and therefore good specific mechanical properties. As such, PMCs are used as light weight, metal replacement materials in many aerospace and automotive applications, for the purpose of reducing vehicle weight, fuel consumption, and ultimately carbon emissions. The other main sectors that PMCs are utilized in are marine, construction, energy generation, and sport and leisure. The following sections give an overview of the different applications in each of these sectors.

- Marine
  Polymer matrix composites, particularly GFRPs, have a long history of use in marine applications particularly for the hulls and decks of yachts, sail boats, fishing boats, lifeboats. Their use reduces the weight of a boat and as such increases performance, speed, acceleration and reduces fuel consumption.

  Over the last 60 years naval vessels have also utilized PMCs for a number of different applications including hulls, mast, shafts, and rudders on various different Frigates and Destroyers. More recently, naval ships such as the Visby Class Corvette and M80 Stiletto are being manufactured using a significant proportion of PMCs. The weight reductions achieved by using composite materials improve range, manoeuvrability and reduce radar reflection for stealth capabilities (Balasubramanian, 2014).
- Aerospace
  Composite materials are used in the aerospace industry for their light weight and excellent mechanical properties. Reducing the weight of an aircraft can increase its speed and manoeuvrability in use and dramatically reduce the fuel consumption over its lifetime. It has been estimated that 4 million litres of fuel can be saved over the 20 year lifetime of an aircraft, when considering the weight saved by using composite materials in its structure (Balasubramanian, 2014). Aircraft body weight reduction also increases the payload an aircraft can carry.

  Similarly to the automotive sector, the first to adopt the use of PMCs as a structural material were those used in high-performance applications, in this case space shuttles and military aircraft (F-14 Tomcat). Although commercial aircraft was slow to follow in such widespread use of PMCs in structural applications, it is now common in new aircraft with the Airbus A380 and Boeing 787 as typical examples. CFRP is now used extensively in structural wing, fuselage, and empennage applications in aircraft throughout the aerospace sector (Balasubramanian, 2014).

The low CTE, low density, and high mechanical properties of CFRPs are resulting in their increased use in spacecraft applications. Some example components are shuttle pressure vessels and support structures for solar arrays and communication devices.

Some examples of military aerospace applications are given in the "Defence" section.

Two of the most common examples of commercial aircraft that used PMCs as a significant proportion of the material used in structural components are the Airbus A380 and Boeing 787 Dreamliner. The A380 has a structural mass consisting of 25% carbon/epoxy composites. The composite components include the empennage stabilizers, wing leading edge, and ailerons and fuselage panels. The Dreamliner has 50% of its structural mass constructed from composites, including the components above and the fuselage and wing box (Gay and Hoa, 2007).

Modern regional aircraft also has around 25% of the structural mass from composite materials and light aircraft such as tourist airplanes and gliders has an even higher percentage of composites in their structure with some gliders constructed almost completely from glass/epoxy composite (Campbell, 2010).

- Automotive

  With the improved performance and efficiency associated with vehicle light-weighting, along with an increased awareness of the environmental impacts of CO_2 emissions, the use of composite materials is growing in the automotive industry. PMCs are utilized in a wide range of applications across a broad spectrum of vehicles on sale today and despite the perception that this is a recent development, their use dates back to the 1950s. In 1955, Citroen manufactured the DS 19 with a glass/polyester composite roof (Gay and Hoa, 2007). In a modern car, some typical components manufactured from PMCs are as follows:

  - engine covers, manufactured from injection molded short glass fiber reinforced thermoplastic

- oil pans, compression molded from SMC
- fuel tanks, manufactured by multilayer extrusion blow molding (see Chapter 1). This is not so obviously defined a PMC, but consider the properties achieved with the laminate construction again and the function of the tie layers.

Some more recently developed composite components found in particular models of car are as follows:

- Volvo XC90 Leaf-springs, manufactured using HP-RTM (this is a variant of RTM using higher pressures) (Reinforced Plastics, 2014)
- A gas-assisted injection molded (see Chapter 4) Audi A8 spare wheel well (Composites World, 2012)
- BMW i3 "life module" or passenger cell manufactured from carbon fiber using RTM (Composites World, 2014)

  As well as the weight saved by manufacturing parts from PMC, the versatility of composite manufacturing methods can result in one complex component replacing an assembly of many smaller metallic parts. This can also save weight by eliminating some fixtures and reduce assembly time.

  CFRP are widely used in motorsport, especially Formula 1; more details are given in the "Sport and Leisure" section.
- Energy generation

  Within the energy generation section PMCs are used to manufacture wind turbine structures and pipes for off-shore oil and gas installations.

  The light weight and high stiffness of continuous fiber composites make them well suited for wind turbines with most structures being manufactured from GRFP composites utilizing epoxies or polyester resins (Chawla, 2012). Composites are typically used to manufacture the blades and nacelle (Thomas et al., 2012) with blades already in the market with CRFP reinforcement replacing GRFP, despite the extra materials costs. Turbine blades are typically made using hand/wet lay-up or resin infusion technique (Campbell, 2010) or filament winding (Gay and Hoa, 2007).
- Construction and infrastructure

  Polymer matrix composites are growing in use in the construction and infrastructure sector replacing or complementing traditional materials in many applications. The driving factors for using composites in this sector are lower in-life maintenance of components and cost and corrosion reduction. Some examples of construction components commonly made using PMCs are reinforced panels, corrugated sheeting, and window/ door frames (Thomas et al., 2012). Large underground gas and sewage pipes are also manufactured using GFRPs (Gay and Hoa, 2007). As well as manufacturing new components using PMCs, existing structures are being retrofitted with supplementary composite material to reinforce existing components. FRPs are utilized in the upgrading and improving the repair of existing bridge columns and other steel-reinforced concrete structures. This is done by wrapping composite (usually a carbon fiber reinforced thermoset) around the concrete pillar (Chawla, 2012). This has also been shown to have seismic benefits (Thomas et al., 2012). Beams can also be retrofitted in a similar way by applying sections of composite. The high stiffness-to-weight ratio of PMCs also means that bridges utilizing them can be constructed with longer spans between supports, bridge decks and overpasses are also structure manufactured using PMCs (Balasubramanian, 2014).

- Sport and leisure
  Composite materials, especially PMCs, are used extensively in the sports and leisure industry because of the excellent lightweight and stiffness properties. Some examples of typical GFRP composite applications are tent poles, hockey sticks, and surfboards. Carbon fiber reinforced PMCs are used in the manufacture of racing bicycle frames, fishing rods and can be used to make musical instruments such as guitars and cellos (Thomas et al., 2012). Golf club shafts can also be manufactured from hybrid carbon and boron fiber composites. Skis are made using a composite sandwich structure (Balasubramanian, 2014) and helmets and canoes are some examples of sporting goods made using discontinuous fiber composites. Motorsport also uses both GFRP and CRFP to make custom one-off components, for example, bodyshell or wing mirror components.
- Defence
  Both hard and soft armor are constructed from PMCs with Kevlar probably being the most well-known brand. Kevlar is an aramid fiber that can be woven into a fabric that, upon impact, deforms providing ballistic protect. Personal protection equipment such as soft body armor incorporates multiple layers of fabrics and hard armor can be manufactured

by combining the fabric with a resin (Chawla, 2012). Polyethylene woven fabrics are also used for manufacturing helmets and soft body armor designed for bullet protection (Balasubramanian, 2014).

As well as for ballistic protection, the defence sector also utilizes PMCs in the construction of naval vessels, military vehicle, and aircraft.

Some examples of military aircraft PMC applications include stealth skins, fighter aircraft airframes, and helicopter and rotor blades. Manufacturing aircraft skins using coated carbon fiber composites can help camouflage an aircraft by decreasing detection by radar reflection and heat emission. Fighter aircraft that are commonly transported on aircraft carriers are required to achieve vertical or short take offs and landings. This is more easily achieved by reducing the aircraft weight by manufacturing the frames from 25% CFRP. Filament wound helicopter rotor blades can be produced near net-shape and can have their vibration characteristics tailored by altering the fiber placement during manufacture. Boron and carbon/boron hybrid composites are more prevalent in military aircraft since their high price is warranted by their superior properties.

- Other

PMCs are also utilized in the chemical industry for the construction of pipes and pressure/reaction vessels, particularly in GFR polyester (Balasubramanian, 2014).

3.5 Research Trends in PMCs

As we have already described so far in the chapter, PMCs are a wide ranging group of materials, differing by the use of various thermoplastic and thermosetting matrices, discontinuous and continuous reinforcements of different types, and many different processing techniques. Due to this range, to scope for composites research is huge and a few of the main themes are described below.

3.5.1 Raw Material and Processing

Faster cure times and better performing resins are an obvious driver, but more specifically OoA chemistries are of particular interest to manufactures looking to simplify and speed up production and remove the need for long consolidations times in the autoclave.

Another challenge associated with many composite manufacturing processes is the lack of automation. Relying heavily on manual processes results in slow manufacturing times and part quality is very much dependent on the skill of the workforce.

One of the challenges of laminates composites is their limited mechanical properties in the *z*-direction. In a bid to overcome this barrier extensive research is being carried out into *z*-direction reinforcement. This comes in the form of 3D weaving, *z*-pinning, and other fiber processes such as knitting. Another area of composite research is the manufacture of net shape reinforcements.

3.5.2 Hybrid Composites

There are a number of different ways in which PMC can be hybridized. Different types of reinforcing fibers can be combined, different fiber architectures (e.g., laminate and random) can be used in a single composite material, or PMCs can be manufactured incorporating both macro- and nanocomposite technology.

A composite material comprising a matrix with at least two different fiber materials is commonly known as "fiber hybrid (composites)" and research into them is driven by cost efficiencies and the need for manufacturing tougher PMCs for improved properties such as impact and fatigue. This is done by combining low elongation (LE) and high elongation (HE) fibers to impart the strength/stiffness properties from one fiber type and the ductile properties from the other, as well as lessening the disadvantages from each. Generally, both the fiber types still have low elongation but the HE fiber elongation is high relative to that of the LE fiber. Fiber hybrids can be structured in a number of different ways. For a laminate structure, the fibers types can be alternated layer by layer (interlayer), bundles of different fibers can be combined together in each ply (intralayer), or different fibers can be intermingled together in the fiber bundles (intrayarn). Combining two different fiber materials can have a hybrid effect; a positive hybrid effect is when the performance of the hybrid composite is better than that predicted using the rule of mixtures.

Extensive research has been carried out over decades into hybrid composites manufactured with carbon and glass reinforcements as well as some studies into hybrids combining carbon with fibers such as aramid, PP, PET, and silicon carbide. It has been found that in UD hybrid composites the tensile modulus generally followed the rule of mixtures, the tensile failure strain had the greatest increase in hybrid effect of 10–50%, and the tensile strength properties of intralayer hybrids were slightly higher than interlayer hybrids. The flexural modulus of fiber hybrids, again, generally follows the rule of mixtures; however, the flexural strength of a hybrid can exceed that of full composite counterparts when intralayer structures are used asymmetrically to optimize the material to the tension–compression loads on each side of the specimen.

More recently, research has been undertaken into using hybrid fibers to manufacture composites with a more gradual failure than is usually seen in FRPs. This is known as pseudoductility and will reduce the catastrophic failure of components and reduce the size of safety factors employed when manufacturing a component with PMCs. Pseudoductility has been achieved in fiber hybrids but only with composites made with a low LE FVF. Another area of fiber hybrid research is in composites constructed using highly ductile fibers as the HE reinforcement, such as polymer and natural fibers. By combining fibers with a greater difference in their failure strains, it is hoped that the hybrid effect could be increased. A more ductile failure has been observed when hybridizing brittle composites with HE fibers and improved tensile and flexural properties have been reported. Natural fiber hybrids are also an area of current research. Combining natural fibers with glass fiber generally offers improved mechanical properties and durability over full natural composites and improved impact properties over full glass counterparts. Hybrids consisting of two natural fibers are studied more for their balance of properties rather than the hybrid effect (Swolfs et al., 2014).

One study has also found a damage-healing potential for hybrid composites. By intermingling glass and PP fibers into yarns, from which a cross-ply NCF was manufactured and made into a composite with an epoxy resin, localized heating was used to repair composite damage caused by an impact. The healed composites achieved a 65% strength increase in compression after impact tests compared to unhealed composite (Selver et al., 2015). Some further research into self-healing composites is described in the multifunctionality section.

3.5.3 Green and Biocomposites

Sustainable technology is a topic that spans all areas of materials research and composites is no exception. PMCs made using natural fibers and/or biopolymer matrices is an established area of research that continues to grow, traditionally investigating the types of fibers and matrices that can be used and more recently looking at improving the performance and processibility of green composites and matching particular materials to the most suitable green composite applications. Although green composites generally have excellent specific stiffness, a number of techniques have been investigated to improve the overall mechanical performance of green polymers; these techniques include modifying the fiber surface to improve adhesion to the matrix, using adhesion promoters in the matrix and developing processing techniques that improve reinforcement dispersion. Common fiber types to be investigated are jute, flax, kenaf, sisal, and coir, and these are typically manufactured into composites with polyolefins or biopolymers such as PLA, PHB, and polysaccharides. Surface treatments such as alkali, silane, peroxide, and anhydrides as chemical options and plasma treatment have been observed to successfully improve fiber surface and wettability. One disadvantage to these treatments is the additional cost to the resulting composite material (La Mantia, Morreale 2011).

The high specific stiffness of green composites lends their use to weight critical applications such as those in the automotive industry. However, one study has shown that there are significant challenges for using green composites in any automotive applications beyond what is already available. Some limitations are the high cost of bioresins, the low microbial resistance of natural fibers, the inconsistency of fiber properties, and the durability of A-class surfaces for the duration of the vehicle life (Koronis et al., 2013).

Despite recent developments in green composite their use is still limited to nonstructural components, packaging, and panels. The reasons for this are the variability in the properties of the reinforcement (inherent due to their natural raw material), lack of fiber alignment in the most commonly manufactured green composites resulting in lower property composites, and the high water absorption and degradation of fibers causing in-life damage to green composite components. An alternative solution to overcoming these challenges is to find applications that complement these properties of green composites. Some examples given in the research include short-life products such as smartphones, PCs, and toys; high specific stiffness sporting goods that provide vibrational dampening such a skis; and biomedical applications where hydrophilicity and biodegradability are an advantage (Dicker et al., 2014).

3.5.4 Joining

As composites are often manufactured to be assembled or fitted onto other components, joining and assembly is a key part of the manufacturing process. Although many joining solutions are proposed in the literature, one of the key challenges that has not yet been met is how to scale up the proposed techniques. In order for PMCs to be used widely in high-volume industries (such as automotive), a high-volume joining process must be found.

Due to the wide applicability of composites, "joining" covers a wide variety of potential materials, processes, and final products. This could involve the joining of similar and/or dissimilar combinations, retrofitting onto long-standing existing structures (e.g., a civil structure) and the solution may be required to be permanent, temporary, or even completely reversible. Given all these issues the production of integrated and multifunctional components provides a potential solution in some cases.

3.5.5 Recycling

Composite recycling challenges can be arbitrarily split between thermoplastic and thermoset resins, each of which presents its own challenges. Preferential recovery of carbon fiber from the polymer matrix (favorable due to cost and environmental considerations) is now done on a commercial scale, and the properties of the recovered carbon fiber remain usefully high for reuse potential. There is, therefore, considerable research effort in Europe currently being expended in exploring the potential in utilizing the recovered fiber.

With composite landfill bans, and the low combustability of highly filled composites for energy recovery techniques, thermoset composites still require technical advancements in expansion of disposal options applicable on a large scale.

Thermoplastic-based materials prevent less of a challenge (see Chapter 4), but mechanical recycling techniques reduce the effectiveness and therefore reduce the market value of long fiber composites. Again, next use applications are needed to take up more recycled materials.

The broader and longer term impacts of the introduction of newer materials such as CNTs into the recycling streams are still not known. The separation (or not) of "green" composites is also a challenge for future waste streams. The impact of biodegradable materials accidently mixed into a nonbiodegradable mechanical recycling-based system and how these may interact and be managed is also not yet clear.

3.5.6 Multifunctionality

The manufacture of multifunctional materials and components is a concept that spans a wide range of research areas, and composites are no exception. The idea of producing a part that has functional abilities in addition to being a structural component of a larger structure appeals from a weight-saving as well as a commercial product enhancement perspective. Some multifunctionality aspects that are the current focus within composite research are damage detection, sensing, self-healing, and electroactive capabilities.

There are a number of ways to add functionality that were introduced in Chapter 1, but the common and simple way of manufacturing multifunctional composite is by loading the polymers with some sort of active additive. The characteristics typically imparted to a polymer using such fillers are varied as can be seen in Chapter 2 but could be to add conductivity, magnetism, or sensing capability to name but three possibilities.

3.5.6.1 Sensing

Nondestructive testing and damage detection for lifetime/in-service health monitoring of PMC components is another interesting area of current research. Fatigue and impact conditions can result in microcrack damage within the composite material resulting in a weakened structure. The ability to monitor, detect, and even notify of or respond to this damage could significantly reduce the likelihood of absolute failure of in-service components.

Monitoring in composites has been achieved using individual embedded fibers or yarns acting as sensors. Nair et al. (2015) manufactured composite materials incorporating multimeasurand optical fiber sensors and these and similar research has been used to detect and measure various manufactured component properties. Process monitoring has also been achieved by using the optical fiber sensor to detect parameters such as temperature, resin refractive index, and strain. Optical fiber sensors have been used for real-life damage detection in existing civil structures such as bridges.

Another study of composite monitoring used piezo-resistive exPAN carbon filament as strain sensors in continuous GFR thermoplastic. It was investigated as a technique in woven, knitted, and tailored fiber placement fabric composites; the yarn was interlaced throughout the fabric and a distortion of the yarn causes a change in its resistance that can be detected.

Another technique investigated for detecting damage within composite components is the use of hollow glass fibers containing detectable liquids such as ultraviolet fluorescent dye. Epoxy composites were manufactured using a hollow glass fiber fabric filled with the UV dye, these were impact damaged and the released UV dye was observed in the damaged areas. Although this technique successfully showed where the composite damage was, it is limited to applications with surface damage on components that are visible when in use.

A significant area of PMC research is around the concept of self-healing composites. The idea that a composite material has the inherent ability to repair any damage that occurs during its lifetime is a very appealing notion. Some healing techniques currently under research are in hybrid thermoplastic–thermoset composite materials, composites manufactured using intrinsically self-healing polymer matrices and composites containing encapsulated uncured resin.

3.5.6.2 Encapsulated Resin/Accelerator – no External Stimulus Required

The study conducted by Kling and Czigany (2014) used hollow glass fibers as a composite damage detection method and also used the same technique as a self-repair technique. This was achieved by filling the hollow fibers with polyester resin and accelerator. GFR epoxy composites were manufactured using a hollow glass fiber (HGF) fabric using hand lay-up and VARTM techniques. Unfilled and resin-filled composites were three-point bend tested before impact damage, after damage, and following two different healing regimes. The healed hand lay-up composites had 23% and 20% higher flexural strength and modulus compared to the damaged composite; the equivalent VARTM results were 25% and 46%.

As mentioned in the "Hybrid composites" section, one technique for manufacturing self-healing composites is to incorporate thermoplastic fibers into the composite structure, such as the Selver et al. (2015) example given earlier. Another method is for thermoplastic particles to be dispersed among the thermoset matrix. The thermoplastic materials used in this self-healing process must have some specific properties, such as low melt temperature and viscosity, and functional groups that can react with or have good adhesion to thermosets. The healing potential of EMMA, PEGMA, and EVA has been investigated, comparing the tensile and compressive properties undamaged, damaged, and repaired coupons manufactured using each thermoplastic healing agent and unmodified epoxy-based composite coupons. While the presence of the healing agents reduces the tensile and compressive properties of the undamaged coupons, the healing recovery of compressive properties ranged from 20% to14%. The tensile healing recovery was more variable, with values ranging from −7.8% to 12% (Pingkarawat et al., 2014).

Much greater healing efficiencies are achieved by composites manufactured using intrinsically self-healing polymer (ISHP) matrices; however, the scale of damaged that can be repaired using this mechanism is much smaller than that of resin-encapsulated of thermoplastic-hybrid composites. Intrinsically, healing polymers contain reversible chemical bonds that can re-form an unlimited number of times; however, an external stimuli is needed to initiate the bond repairing chemical reaction. Some typical stimuli used are heat, moisture, irradiation, and electrical. Structural composites manufactured using ISHPs have been reported to have achieved

41%, >90%, and even 100% strength healing efficiencies and good stiffness recovery in short fiber PMCs.

Functional composites comprising actively filled ISHPs have also been investigated. The high recovery efficiencies of electrical thermal conductivity, EMI shielding effect, magnetic, and electro-mechanical properties have all been reported (Zhong and Post, 2015).

3.5.6.3 Piezoelectric/electroactive Systems

Electroactive composites introduce the concept of electrical functionality into composite structures that feeds an active area of research into energy harvesting, energy storage, and embedding electrical functionality into the structure, thereby removing the need for wiring harnesses. A composite structure that can harvest and transfer the energy from its basic function (such as a solar cell) would remove entirely the need for energy input. For industries such as marine, aerospace, military, and automotive there are obviously considerable advantages in such self-powered concepts. These kinds of systems require a range of integrated functions and are therefore considerably more complex in conception than, for example, self-sensing composites. Embedding all the materials within the polymer matrix material is unlikely and such systems are far more likely to be a complex assembly of individual laminates that all build into a final connected structure. Glimpses of this future can be seen in individual material and processing advances (also seen in Chapter 4), but we are yet to assemble a fully multifunctioning complex composite as yet.

3.5.6.4 Multifunctionality of Other Composite Materials and Potential

This chapter has covered just a small section of the composite area focusing on PMC; there is much active research in other areas of composite materials science be it from graphene to ceramics, research in metal matrix composites (MMC), and wood and natural composites all providing alternative properties, costs, and solutions. Opportunities for these areas can in time also provide future opportunities to PMC due to the versatility of PMCs and therefore ability to accommodate a wide variety of other materials directly as formulation or as hybrid structures and products.

One interesting development is the creation of the "Bucky syntactic foam" that was reported by Ghamsari in 2014 (Ghamsari et al., 2014) Carbon nanotube–ionic liquid hybrid was used to hollow "balloons" that were embedded within a polymer matrix. The use of these materials created a conductive flow path, creating a conductive rather than insulating electrical field. These materials are predicted by the authors to have a wide field of application potential from electronic packaging, structural health monitoring and to provide EMI-shielding to structural composites.

Therefore, the expansive nature in exploiting multifunctional design is both awe inspiring and considerable. In order to understand what hold such developments back, a look at some of the barriers is therefore necessary. This should also be considered with the barriers identified in Chapter 4 in mind.

3.6 Barriers to Multifunctionality

The incorporation of new materials into existing formulations, especially for long-term applications, requires an extensive regime of testing to ensure safety of applications. As the complexity of any composite increases the ability to model and predict the entire system performance through its performance life becomes increasingly complex as does the potential decommissioning and disposal of such a complex component. Therefore, introductions of products with advanced functionality are not without complexity and therefore require a broad and strategic view of impacts.

3.6.1 Financial Impact

While cost does not often factor into blue sky research, it is without doubt the major consideration for a manufacture. No product will survive without a sound profit/cost model. Comparable fiber costs are given in Table 3.1 and the cost implications in using expensive materials can be clearly seen with reinforcement and can equally be applied to the addition of other materials. For functionality be it of a metallic, polymeric, and ceramic nature, it is unlikely to add up economically without the offsets derived from the elimination of other components and manufacturing routes. System integration of some kind is, therefore, the model recognized in mass volume industries such as injection molding (Chapter 4). For composites this question is more complex due to the "cottage" nature of lots of composite manufacturing but perhaps where these two industries most overlap is in the automotive industry that has embraced a process integration ethic for many of their components.

3.6.2 Political Impact

Politics can have a big impact on the economy and the priorities given to science and manufacturing in individual countries play a major part in the economic climate that manufacturers must work in and the support for research and development.

Countries such as Germany have long understood this; other countries such as the UK are catching up with continued support from government in creating and sustaining Catapult centers to support UK manufacturing industry with most relevance to this chapter being for High Volume Manufacturing (HVM) based at WMG, University of Warwick, and the National Composites Center (NCC) based in Bristol.

3.6.3 Environmental Impacts

Legislative drivers, especially in the European Union, are pushing manufacturers toward greater producer responsibility for the entire lifecycle of their products and to consider an entire cradle to grave perspective (see Chapter 4 for more details). Specifically for the composites industry in December 2013, "Composite Materials: A Resource Efficiency Action Plan" was written on behalf of the UK Composites Industry Stakeholders Group in order to identify a series of potential actions that would reduce waste and improve material efficiency in the UK composites sector (UK Composites Stakeholder Group, 2013). The report focuses on GFRP since the production and waste volumes of composite manufactured with this type of reinforcement far outweigh any other type. It identified that many of the factors driving the growth in the UK composites market are due to EU legislation often associated with the end-of-life of products.

The UK Landfill Tax also impacts waste management and therefore is an important consideration for the composites industry. The standard tax for disposing one tonne of waste into landfill from 2014 is £80, but in reality it is likely to be much greater than this due to additional fees and costs. This acts as a significant motivator to develop alternative waste management.

The Waste Electrical and Electronic Equipment (WEEE) directive states that WEEE manufacturers must pay costs including some associated with the recyclability of products. It is intended to encouraging manufacturers to manufacture products that are more easily recycled and as such may affect the types of composites used in the manufacture of WEEE products.

The UK has End-of-Life Vehicle (ELV) Regulations, based on the EU ELV Directive, that declare that 95% of vehicle landfill waste must be redirected to recycling (85%) and energy recovery (10%). Therefore, if composites are to replace metals as a lightweight alternative in automotive applications, there must be a suitable composite recycling route comparable to that of the current one for metals.

Other drivers for improving the recyclability within the composites industry include corporate social responsibility and environmental accreditations. Organizations often pledge commitments to develop more environmentally responsible processes and seek to produce products that comply with particular accreditations to appeal to customers/the public and acquire leading position in the market.

The aim of the report is to increase the composite industries efficiency by improving categorization of composite waste, support new recycling methods, developing composite recyclate markets, and encouraging the dissemination of best practice. This will be achieved by developing business models for recycling and case studies of recyclate applications, facilitating cooperation between industry and universities to develop novel processes. Waste streams would be evaluated and ID labeling investigated as well as developing a waste grading system. A database of composites data for life cycle analysis will also be considered.

3.6.4 Social Impacts

The major social impact is one of wealth creation through more advanced technologies and products providing more manufacturing jobs; however, this needs to be done without detriment to the living environment in a way that affects longer term human health. Multifunctional composites provide great potential to enhance all aspects of our every day lives from transport systems to buildings and beyond and also to monitor and protect this infrastructure for our benefit.

3.7 Conclusions: Advantages and Disadvantages

It has been shown that polymer matrix composites (PMCs) are used extensively in many wide ranging applications that have been described in this chapter. There is considerable research being carried out into a number of different areas of composite technology and as a group of materials; PMCs are often hailed as the solution to a lot of problems. However, there are also considerable challenges associated with complex material structures along with the heralded benefits long associated with replacing traditional or metallic materials with composites. The following sections summarize the main advantages and disadvantages to using multifunctional PMCs.

3.7.1 Advantages

- Low processing and tooling costs (compared to metal)
- Lightweight

- Process integration potential
- Tailored properties
- Corrosion resistance
- Recycling

3.7.2 Disadvantages

- Difficult to repair
- Difficult to predict properties
- Difficult to model properties
- High material cost
- Joining and assembly
- Lack of automation
- Recycling

That recycling appears on both lists relates to the broad nature of PMCs (thermoset/thermoplastic/loading levels) and the added complexity that multifunctionality brings to disposal. However, what should be clear is whatever the strategy of functionalizing, there is no doubt considerable potential to be exploited.

Abbreviations

BMC	Bulk molding compound
CF	Carbon fiber
CFRP	Carbon fiber reinforced plastic/polymer
CNT	Carbon nanotubes
CSM	Chopped strand mat
ELV	End of life vehicle
EMI	Electromagnetic interference
EMMA	Ethylene/methyl methacrylate
EVA	Ethylene-vinyl acetate
FRP	Fiber reinforced polymers (plastics)
FRTPs	Fiber reinforced thermoplastics
FVF	Fiber volume fraction
GF	Glass fiber
GFRP	Glass fiber reinforced plastic/polymer
GMT	Glass mat thermoplastic
HE	High elongation
HGF	Hollow glass fiber
HM	High modulus
HP-RTM	High-pressure resin transfer molding

HS	High strength
HVM	High-volume manufacturing
IM	Intermediate modulus
ISHP	Intrinsically self-healing polymers
LCA	Life cycle analysis
LE	Low elongation
MMC	Metal matric composite
NCF	Non crimp fabric
OoA	Out of autoclave
PA	Polyamide
PAN	Polyacrylonitrile
PEEK	Polyether ether ketone
PEGMA	Poly(ethylene glycol) monomethyl ether methacylate
PEI	Polyetherimide
PET	Polyethylene terephthalate
PHB	Polyhydroxybutyrate
PLA	Polylactic acid
PMC	Polymer matrix composites
PP	Polypropylene
PPS	Polyphenylene sulfide
Pre-preg	Pre-impregnated fabric
RIM	Reaction/resin injection molding
RRIM	Reinforced reaction/resin injection molding
RTM	Resin transfer molding
SMC	Sheet molding compound
SRIM	Structural reaction/resin injection molding
SRP	Self-reinforced plastic/polymer
TP	Thermoplastic
TS	Thermoset
UD	Uni-directional
UHM	Ultrahigh modulus
VARTM	Vacuum-assisted resin transfer molding
VBO	Vac-bag only
WEEE	Waste electrical and electronic equipment

References

Balasubramanian, M., 2014. Composite Materials and Processing. CRC Press, Boca Raton, FL.

Bishop, G., Halliwell, S., Mafeld, A., 2009. The UK carbon fibre composites industry: market profile. Department for Business, Innovation and Skills, London.

Campbell, F.C., 2010. Structural Composite Materials. ASM International, Ohio.
Chawla, K.K., 2012. Composite Materials: Science and Engineering, third ed. Springer, New York.
Composites World. 2012. Spare wheel well: functional integration. [Online] Available from: http://www.compositesworld.com/articles/spare-wheel-well-functional-integration (accessed 29.01.15.).
Composites World. 2014. BMW Leipzig: The epicenter of i3 production. [Online] Available from: http://www.compositesworld.com/articles/bmw-leipzig-the-epicenter-of-i3-production (accessed 29.01.15.).
Dicker, M.P.M., et al., 2014. Green composites: a review of material attributes and complementary applications. Compos. Part A 56, 280–289.
Gay, D., Hoa, S.V., 2007. Composite Materials: Design and Applications, second ed. CRC Press, Boca Raton.
Ghamsari, A.K., Wicker, S., Woldesenbet, E., 2014. Bucky syntactic foam; multi-functional composite utilizing carbon nanotubes-ionic liquid hybrid. Compos. Part B 67, 1–8.
Kling, S., Czigany, T., 2014. Damage detection and self-repair in hollow glass fibre fabric-reinforced epoxy composites via fibre filling. Compos. Sci. Technol. 99, 82–99.
Koronis, G., SilVA, A., Fontul, M., 2013. Green composites: a review of adequate materials for automotive applications. Compos. Part B 44, 120–127.
Morreale, M., LA Mantia, F.P., 2011. Green composites: a brief review. Compos. Part A 42, 579–588.
Nair, A.K., et al., 2015. Process monitoring of fibre reinforced composites using a multi-measurand fibre-optic sensor. Sensors Actuators B Chem. 212, 93–106.
Pingkarawat, K., et al., 2014. Mechanical properties of mendable composites containing self-healing thermoplastic agents. Compos. Part A 65, 10–18.
Reinforced plastics, 2014. Volvo XC90 features polyurethane composite leaf spring. [Online] Available from: http://www.reinforcedplastics.com/view/40293/volvo-xc90-features-polyurethane-composite-leaf-spring/ (accessed 29.01.15.).
Selver, E., et al., 2015. Healing potential of hybrid materials for structural composites. Compos. Structures 122, 57–66.
Swolfs, Y., Gorbatikh, L., Verpoest, I., 2014. Fibre hybridisation in polymer composites: a review. Compos. Part A 67, 181–200.
Thomas, S. et al.,(Ed.), 2012. Macro- and Microcomposites. Polymer Composites vol. 1. Wiley-VCH, Weinheim.
United Kingdom. UK Composites Industry Stakeholder Group, 2013. Composite Materials: A Resource Efficiency Action Plan. The Green Construction Board, UK, (Report 021).
United Kingdom, UK Trade and Investment, 2010. UK Composites Supply Chain Study. HMSO, London.
Zhong, N., Post, W., 2015. Self-repair of structural and functional composites with intrinsically self-healing polymer matrices: a review. Compos. Part A 69, 226–239.

Further Reading

Chou, T.W. (Ed.), 1993. Structure and properties of composites. Materials Science and Technology, vol. 13. Weinheim, VCH.

Chung, D.L., 2010. Composite materials: science and applications, second ed. Engineering Materials and Processes Springer, London.

Hull, D., Clyne, T.W., 1996. An Introduction to Composite Materials. Cambridge Solid State Science Series, second ed. Cambridge University Press, Cambridge.

Kaw, A.K., 2006. Mechanics of composite materials, second ed. Mechanical Engineering Series CRC Press, Boca Raton.

Nicolais, L., Meo, M., Milella, E. (Eds.), 2011. Composite Materials: A Vision for the Future. Springer, London.

United Kingdom. Department for Business, Innovation and Skills, 2009. The UK Composites Strategy. HMSO, London.

Witten, E., Jahn, B., 2013. Composites Market Report 2013. AVK, Germany.

4 Injection Molding of Thermoplastics

Vannessa Goodship

4.1 Introduction

Injection molding was invented in 1872 and is a well-established and well-developed production method for producing a wide variety of complex and identical mass produced goods. Using variations on the basic injection molding process mean it can be used to mold both thermoplastic and thermoset materials, as well as other materials such as liquid silicone rubber (LSR) and metal powder. The underlying fundamental machine design will vary dependent upon which of these specific material types is used but the basic principles of inject material, form material, remove formed component remain throughout.

By far the most widely used materials for injection molding are the thermoplastics. A typical thermoplastic injection molding machine is shown in Figure 4.1. Materials such as high density polyethylene (HDPE), low density polyethylene (LDPE), and polypropylene (PP) are the most widely used materials in this process for the production of high volume consumer goods such as closures: hinged bottle tops, clips, lids, and pots. Examples of these kinds of products are shown in Figure 4.2. Therefore, this chapter will mainly but not exclusively focus upon thermoplastic materials as the substrates on which to base a later discussion on multifunctionality. For other material types the reader will be referred to specific sections or the further reading section of this chapter.

Like other thermoplastic processes such as extrusion, blow molding, rotational molding, and calendaring, injection molding can be thought of as two very simple process steps:

1. Make the shape – by polymer melting or softening and then shaping
2. Fix the shape – by cooling the polymer

This chapter will expand considerably on this accurate, but simplistic, concept and look at the machinery and wider polymer and processing interrelationships that allow injection molding to be such a successful, flexible, and widely used commercial processing route.

Design and Manufacture of Plastic Components for Multifunctionality
http://dx.doi.org/10.1016/B978-0-323-34061-8.00004-1

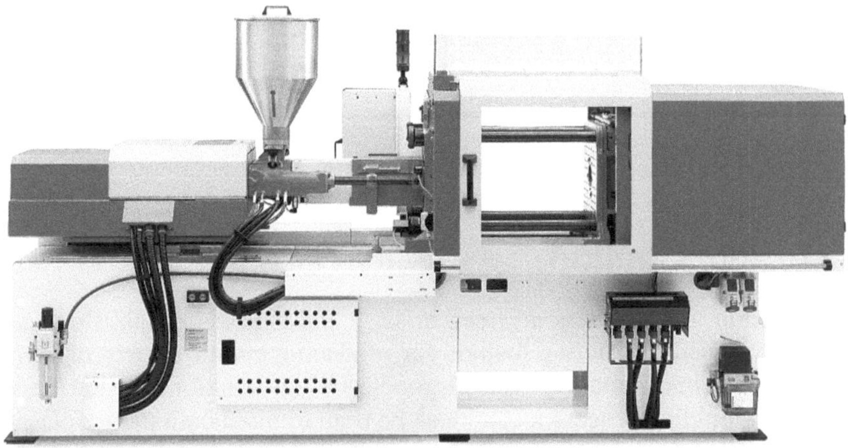

Figure 4.1 Injection molding machine. PHOTO: ARBURG.

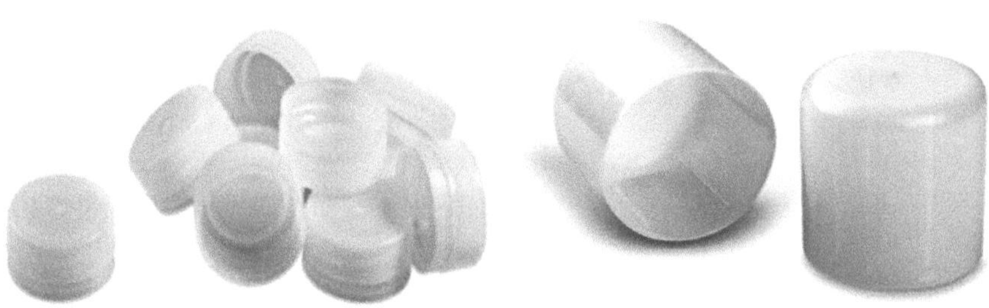

Figure 4.2 Examples of closures produced by injection molding.

It will also consider the injection molding process in depth and illustrate machine design related to both initial material types and the final desired outcome. It will also discuss periphery equipment that can be added onto the standard injection molding process and slowly develop understanding of multifunctional aspects through a variety of examples of increasing design complexity. One of the greatest strengths of injection molding technology is the ability to integrate other processes into the injection molding production cycle to produce a more complex component. As a common example of this, consider the injection molding of a screw driver handle. The process includes robotic placement of the premanufactured metal driver between each cycle before the plastic handle is injected and shrinks around it. This results in a product incorporating both metal and plastic by the process overmolding.

Overmolding is also used to produce multiple color effects of the same material in different colors as displayed by the vast array of toothbrush

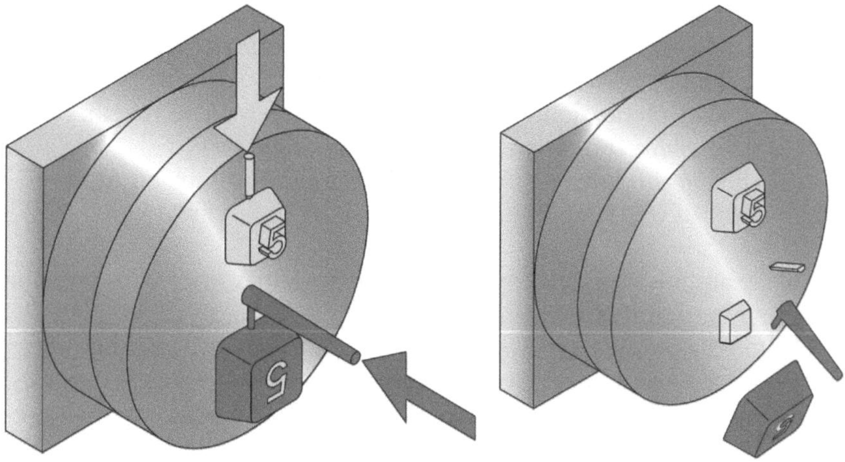

Figure 4.3 Overmolding to produce a two-colored component. Illustration courtesy of ARBURG.

designs that are readily available. In this way the range of products that can be produced by injection molding is expanded. Figure 4.3 illustrates the concept of overmolding to produce a two-colored component. This example combines two thermoplastic materials together to create a numbered keyboard switch.

Overmolding and other process variations will be introduced in more depth in Section 4.5, but can produce considerable cost savings when used in suitable components as Table 4.1 illustrates.

The next example compares the costs of three different production routes for a knob for an electrical device.

Table 4.1 Comparative Costs of Integrated Production Methods

Production Method	Traditional Cost	2-Shot Cost	3-Shot Cost
Tool cost (£)	8500	12,000	19,000
Cycle time (s)	21 + 21	21	21
Machine cost (100,000 pieces)	1241	657	695
Materials	345	365	385
Assembly/print	1650	825	0
Job cost	3236	1847	1080

1. Traditional production: A knob and a colored cap are both molded separately. They are then assembled together and printed.
2. 2-shot production: The knob and cap are molded together and then a further printing process is undertaken.
3. 3-shot production: The print effects are also molded in as part of overall design.

This example shows clearly some of the tradeoffs to be considered in designing for multifunctional components.

The tooling cost is increased as the part complexity increases; however, there were cost gains in the fact, only one tool instead of two was now used in production, and therefore only one injection machine not two was required. The use of an indexing plate (see Section 4.4) meant that three cavities were very effectively being used for the production process at any one time to build the component up shot by shot, so the cycle time in this case did not increase. The final consideration is the cost gains achieved by the removal of the entire costs and time associated with assembly and printing.

As plastic processing has developed around the intrinsic properties of the polymeric materials themselves namely their low thermal conductivity and processing temperatures and stability, this chapter will now begin by stepping back to look at the basic types and properties of polymer materials, the feedstock of plastic injection molding process.

4.2 Polymer Materials

4.2.1 Classification

Polymer materials on their own cannot generally be used for injection molding. They need to be modified with additives (usually integrated into production by the polymer manufacturer) to give suitable light resistance, heat resistance, and useful physical properties. These materials can then be further modified if needed by compounding processes, that is, extrusion, to give additional function such as reinforcement, specific color, functional additives, etc. to supply a suitable plastic feedstock material. Additives will be discussed later in this section, for now the different classes of polymers are considered.

There are three distinct types of classes of plastic material that will be discussed in this section namely:

Thermoplastic: materials can be remelted and reprocessed (e.g., polypropylene)

Thermoset: chemically cross-linked materials that cannot be reprocessed (e.g., epoxy resin)

Elastomer: can be either thermoset or thermoplastic, but has the property of being able to be repeatedly stretched but still return to its original shape repeatedly (e.g., rubber)

There is a further need in light of trends in new materials and environmental legislation to clarify and distinguish between the origin of the feedstock material that can effect sustainability, life cycle, recyclability, and degradation decisions (these are further covered in Section 4.6).

Bio-based polymers are derived from biomass rather than crude oil, but can still be further split into synthetic bio-based polymers and natural bio-based polymers.

As a common example polypropylene can be produced from both oil and sugarbeet, but it is still synthetic polypropylene that is produced. It is just the raw material source that is different, one being a bio base and one being crude oil base. Therefore, bio-based plastics are not necessarily degradable or naturally occurring. Bio-based plastics are measured by their origin (C_{14} content as the renewable carbon measure) not their necessity to degrade.

To be degradable a material will break down after a specified period of time into smaller constituents when exposed to a particular environment such as heat, oxygen, visible light, or UV light. This can occur with both synthetic and natural polymers mainly through the use of additive or blends (combination of polymers) technology. There is also a synthetic polymer that dissolves in water, namely Polyvinyl alcohol (PVOH). This is the material used for water liquid capsules, which dissolve in the machine to release the liquid. This is a completely synthetic polymer.

Biodegradation refers to the specific action of microorganisms in bringing about the breakdown into smaller constituent units and raw materials and the decomposition into carbon dioxide, biomass, and water.

Finally, compostable plastics must meet the compostability standard EN 13432, which lays down standards relating to the time taken to achieve a set level of degradation. It should be noted this standard relates to particular products and not the polymers themselves as compostability depends on the design of the component such as weight and wall thickness for instance.

Thermoplastic materials are also commercially available as second use recyclate materials. Not surprisingly thermoset materials are not, though powdered thermoset materials can be used as filler materials (see the additives section).

4.2.2 Thermosets

Thermosets are materials that undergo chemical changes to their structure during the injection molding process. They go from being essentially thermoplastic in form and nature before molding to very stiff and brittle to a point where they cannot be melted again. This process is called curing, and happens as chemical bonds form between the polymeric chains forming a highly cross-linked network that cannot flow or be broken again by the further addition of heat. A thermoset material will degrade and burn at high temperatures but will not melt. These materials are used where strength, durability, and high heat resistance are required. They are also often used in conjunction with further reinforcing such as glass or carbon fibers to form incredibly strong composite materials with a range of applications such as wind turbine blades, ship hulls, aeroplane wings, cooker knobs. Examples of generic thermoset materials are shown in Table 4.2.

4.2.3 Elastomers

An elastomer is defined as "a material that at room temperature can be stretched repeatedly to at least twice its original length, and upon release

Table 4.2 Thermoset Materials and Applications With Some Application Examples

Thermoset Polymer	Example Application
Epoxies	Adhesives, insulation for covering/encapsulating electrical components, laminates.
Urea formaldehyde (UF)	Toilet seats, knobs, electrical components
Melamine formaldehyde (MF)	Worktops (laminates and coatings), knobs, handles, electrical breakers.
Phenolics	Not generally used as much as in past (Bakelite) as there are modern alternatives, but specific applications such as brake cylinders.
Polyurethanes	Rigid foams for upholstery and insulation
Unsaturated polyesters	Generally uses with fiber reinforcement such as fiberglass. Auto body filler, satellite dishes, partitions, wind turbine blades, tool housings.
Polyimides	Uses in high wear applications. Bearings, gears also used in microelectronics.

Table 4.3 Some Commonly Used Commercial Elastomer Designations

Designation	Description	Example Subgroups
TPE-A	Polyamide	PA 12, PA 6 base
TPE-E	Polyester	Polyetherester, polyesterester
TPE-O	Polyolefin blends	PP/EDPM (can also be cross-linked)
TPE-S	Styrene	SEBS, SBS
TPE-U	Polyurethane	Polyester urethane, polyether urethane
TPE-V	Thermoplastic vulcanizates. Have gone through vulcanization process.	Various

of the stretch, will return with force to its approximate original length." American Society for Testing and Materials (ASTM)

Elastomers come in both thermoset and thermoplastic varieties. Natural rubber (polyisoprene) is a naturally occurring thermoset rubber material; however, as already discussed thermosets cannot be remelted and therefore recycled. As sustainability and recycling have become of greater developmental interest a range of thermoplastic elastomers (TPE) have been developed to enable the replacement of thermoset (therefore nonrecyclable) rubber-based materials. They can also impart other advantages over thermoset elastomers such as faster cycle times, lower density, and therefore lower weight parts, and they can also be aesthetically marked and colored.

In terms of more advanced processes, TPEs can be particularly useful for overmolding applications when combined with thermoplastics to produce hard/rubber combinations allowing application in handles, shoes heels and soles and handles and grips, as well as more conventional rubber applications such as seals, gaskets, and stoppers. Some common TPE-based materials are shown in Table 4.3.

Thermoplastic elastomers tend to be categorized on two properties not normally generally encountered in thermoplastic applications. One is the compression set that is measured as a percentage of recovery from a set load. This guides the specification of materials required to undergo specific load deformation and likely recovery and lifetime performance at these specified loads, times, and temperatures. 0% indicates a full elastic recovery, 100% indicates no recovery at all. The other useful property indicator for elastomeric application is the shore hardness, which is the resistance to indentation (hardness). There are a number of shore hardness scales, soft rubbery TPEs tend to be measured on the Shore A scale, as these become less rubbery and more rigid they move back toward more rigid behavior of general thermoplastic and thermoset materials that are measured on Shore D hardness. For example, PP-based elastomers can be purchased that are very rubbery (i.e., Shore A 40) but can also be blended with base PP materials to create a complete range of behavior.

4.2.4 Silicone-Based Elastomers

A further type of elastomer, silicone rubber, tends to be covered in more specialized rubber publications. However, its relevance to multifunctional application makes it worth including here. In contrast to the carbon–carbon backbones found in other thermoset and thermoplastic plastics and elastomers plastics, silicones have an unusual chemistry of alternating silicone and oxygen atoms.

This means that is not sourced from oil feedstock like conventional rubbers but is actually made of quartz powder (SiO_2). Liquid silicone rubber (LSR) can be vulcanized in the molding process to make silicone rubber and is used for a wide range of items such as seals and flexible pipes and hoses in the medical industry. Silicone molds are easy to make and widely available and are commonly used to make brightly colored and flexible baking wares, ice cube trays and can also be integrated with other materials in an overall assembly. Some commonly seen examples are shown in Figure 4.4.

4.2.5 Thermoplastics

These materials are able to go through multiple cycles of heating, melting, flow, and cooling again. Therefore, they can be recycled. They make

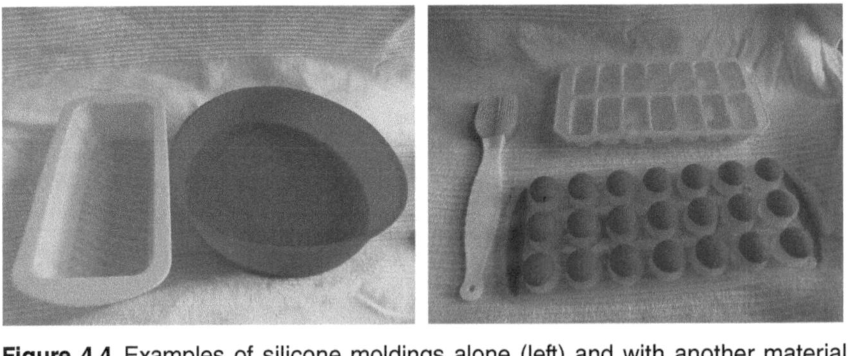

Figure 4.4 Examples of silicone moldings alone (left) and with another material component (right).

Table 4.4 Common Thermoplastic Materials and Typical Applications

Thermoplastic Polymer	Typical Applications	Amorphous or Crystalline
High-density polyethylene	Packaging, crates, wheelie bins	Semicrystalline
Low-density polyethylene	Packaging, grocery bags, toys, lids	Semicrystalline
Linear low-density polyethylene		Semicrystalline
Polypropylene	Yogurt pots, caps and closures, battery casings	Semicrystalline
Polyamide	Bearings, skates wheels, fishing lines	Semicrystalline
Polyethylene terephthalate (PET)	Carbonated drinks bottles	Amorphous
Polyvinyl chloride (PVC)	Packaging, shoes, flooring, window frames	Amorphous
Polystyrene (PS)	Rulers, pens, CD cases	Amorphous

up the vast majority of the volume plastic market, familiar common everyday plastics used in everyday consumer items and food packaging such as soft drinks bottles, carrier bags, bowls, buckets, and wheelie bins. The most common materials are, therefore, often termed the commodity plastics that include common materials such as polypropylene (PP), high density polyethyles (HDPE), polystyrene (PS), polyethylene terephthalate (PET). Examples and applications are shown in Table 4.4.

On cooling thermoplastic materials can have different structures, for example they can be amorphous, crystalline, or a mixture of the two states as shown in Figure 4.5.

4.2.6 Polymer Types Relative to Injection Molding

There are a wide number of polymer materials with various formulations available commercially for the injection molding process. For the purposes of simplicity only thermoplastic materials are considered going forward here. To further this basic discussion on materials for injection molding, the material needs to have the following characteristics.

1. Have the ability to be deformed so that the desired component can be produced.

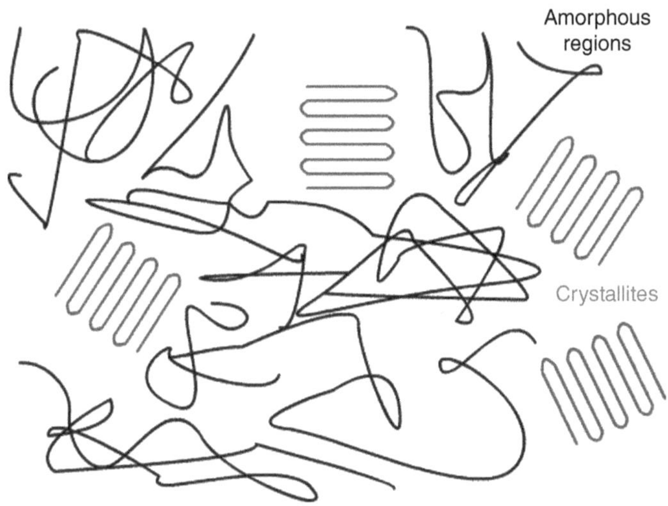

Figure 4.5 Amorphous and semicrystalline structures.

2. Be able to withstand sufficient heat and shear as to allow the feeding, shaping, and cooling process while retaining sufficient useful properties for purpose.
3. Be able to transfer heat in able to solidify again once formed.

During processing and because of the harsh environment of injection molding the material may undergo changes to its structure and properties. Therefore, one of the most important factors in injection molding is the correct setup and programming of the injection molding process settings that vary widely dependent upon the tool, component required, material type, and grade. To get an idea of the kind of harsh environments these materials can be exposed to during the process, a typical material may be exposed to a temperature range anything from 150 to 450°C for several minutes, at barrel pressures from anything 14–205 MPa and be exposed to shear force in the region of 10^5 s^{-1} or higher. Injection molding is the highest shearing process in commercial use for plastic production. Finally, moldings are then stressed even further by cooling the moldings as rapidly as possible for ejection before the start of another production cycle. The cooling cycle has the biggest effect on the economic viability of injection molding; this will be discussed in Section 4.3.7.

Therefore, for every injection molding material there is a specific processing window where satisfactory moldings can be produced time after time for thousands of components. Finding this processing window (see Figure 4.6) for every new injection molding tool and material variation

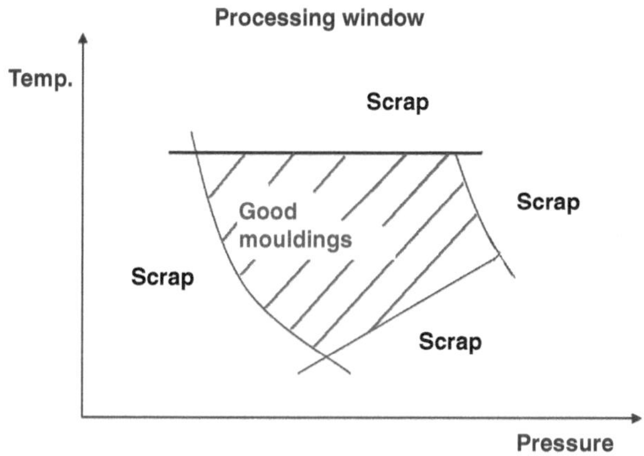

Figure 4.6 Example generic processing window.

is a vital part of setting up a new production run in the injection molding industry.

Material suppliers provide processing guidelines along with data sheets of the expected final performance properties against set standards to manufacturers; these are also widely available on the web for designers. However, understanding the intricate process–property relationships further is important, especially when considering multimaterial systems; therefore, a more in-depth discussion of these factors now follows.

4.2.6.1 Material Appearance

The majority of material for injection molding is received in the form of granules or pea-sized pellets; this makes for easy feeding down through the hopper system and onto the screw, powder material can also be used as can reground plastic. In both cases, if they bridge the feed system it may be necessary to have stirring or a shaker unit in the hopper unit.

4.2.6.2 Amorphous and Crystalline Structures

Amorphous materials soften over a broad temperature range and have a totally disordered and random molecular arrangement as illustrated in Figure 4.5. On cooling therefore this state of randomness remains, and gives the useful property of transparency that amorphous materials possess, the softening allows the chains to become mobile, under the region known as the glass transition (Tg) the chains are frozen in like glass. Crystalline materials, by comparison, tend to be opaque due to highly ordered crystalline regions through which light cannot penetrate. Crystalline structures are present below the melting point (Tm); on melting the polymer takes on a more amorphous-like structure as polymer chains can move. The cooling has a critical role in the development of the crystalline structures, as it depends on the alignment of the chains, how the crystalline structures, and when the temperature is cooled below the Tm. This therefore also impacts shrinkage across the molding. Crystals in a semicrystalline structure impart properties not present in an amorphous structures such as fatigue resistance, resistance to chemical attack, and environmental stress cracking which are all seen in amorphous materials. It should be noted that thermosets before crosslinking and curing can also be classed in this manner.

4.2.6.3 Thermal and Thermal Stability

It is necessary to heat polymer material in order to make it soften and flow. Crystalline polymers have a sharp melting point (Tm) where they turn from solid to liquid; amorphous polymers have a softening range above the glass transition temperature (Tg). Semicrystalline polymers are a mixture of both crystalline and amorphous structures. Some examples of these temperatures are shown in Table 4.5. Once the material is viscous, the polymer molecules can begin to move and rearrange in response to the force that is exerted on them.

Once viscous the viscosity of these materials is material dependent but compared to water (10^{-1} Pa·S) can be thought of as varying from a consistency between "honey" to "modeling clay" generally in the range 2–300 Pa·S, depending on which material is being considered.

A common term used to describe the molten property of a polymer in injection molding is the melt flow index (MFI); this relates to the specific flow rate at a temperature and applied load, for example a common MFI

Table 4.5 Indication of Melting Points and Glass Transitions of Some Generic Polymers

Polymer	A or C	Tm	Tg
Polyethylene (PE)	Semicrystalline	135	−125
Polypropylene (PP)	Semicrystalline	170	−20
Polycarbonate (PC)	Amorphous	–	149
Polyethylene terephthalate (PET)	Semicrystalline	245	69
Polyamide 6 (PA6)	Semicrystalline	233	75
Polymaide 66 (PA66)	Semicrystalline	To add	To add
Polystyrene	Amorphous	–	110

range in injection molding for polypropylene is between 2 and 50. This would generally be written referring to both the temperature and weight the material was measured at, for example 3.5 g/10 min at 220°C, 2.16 kg. This refers to a test carried out at 220°C using one of the standard weights, in this case 2.16 kg. The 3.5 gives an indication as to the flow, 2 can be considered fairly thick and viscous, and 50 very free flowing in relation to a guide for injection molding processing.

Once a load is applied onto the melt then it is the entanglement of individual polymer chains with each other what gives the polymer a viscosity greater than water for example. As individual chains become more and more orientated, this viscosity will lessen. Also as the chain lengths of individual chains shorten (and therefore there is less entanglement per chain), the viscosity will also lessen. It should be noted here that as viscosity drops, MFI value rises as MFI measure flow rate.

As the molecular configuration of individual chains can also vary (linear, branched, or their constituent molecules) and therefore their ability to pack closely and orientate, these are also factors that can cause variants in flow behavior under injection molding conditions and are all related to the molecular weight of the polymer; the effect of chain branching is shown in Table 4.6.

The processing properties and final properties of molding are, therefore, highly linked to both the overall molecular weight and the molecular weight distribution.

The direction of applied force also has an effect; this can be very simply considered by imagining a cube and then all possible directions the force

Table 4.6 Influence of Branching on Polyethylene Properties

Polyethylene	Melting Point (°C)	Density (g/cm^3)	Tensile Strength (MPa)	Number and Type of Branches (per 1000 Carbon Atoms)
LDPE	110–120	0.91–0.93	17–26	Long branches
LLDPE	122–124	0.92	13–27	10–30 short branches
HDPE	130–135	0.94–0.97	21–28	4–10 short branches

Figure 4.7 Polymer chains at rest and in shear.

could be applied from on each face. A further important consideration in polymer processing is the shear force (often termed the shear stress), as illustrated in Figure 4.7.

Shear forces cause the chains to orientate and therefore produce an effect similar to a rise in temperature by causing the viscosity to drop. Once the shear force is reduced again, the polymers will relax and try to recover to their original state. However, exposure to high shear rates can cause chain breakage with a permanent reduction in the chain length and viscosity properties. This critical shear rate is a material specific property.

All thermoplastic polymer materials will degrade above a certain temperature due to a lack of heat stability. It is important processing is carried out within this heat stability range but also with consideration to the total exposure time, as this will also be time dependent to total heat exposure. In relation to dwell time in the injection molding barrel during the injection molding process, this is referred to either as the material dwell time or more commonly the residence time.

In general injection molding polymer materials are very bad at conducting heat; this has meant that polymer processing equipment has had to be designed to maximize heat transfer. Therefore, generally with any processing equipment polymer thicknesses at any point in processing tend to be kept as thin as possible or mixed as well as possible to prevent temperature gradients occurring. This is the reason machines such as extruders and injection molders need reciprocating screw designs in order to maximize the opportunities for heat transfer within the melt. This is also why you tend to see polymer applications where the material is used as an insulator rather than as a conductor like metal materials. There are specialized conductive polymers, which are discussed in Chapter 2, and polymers can be made conductive by the use of other fillers and additives, but generally polymers are thought of as insulating materials as shown in Table 4.7. This has implications for heat transfer when cooling and solidifying as covered in Section 4.3.

Table 4.7 Comparator Thermal Conductivity Values for Common Materials

Material	Thermal Conductivity (W/mK)
Air	0.03
Water	0.6
Metal – steel	35
Metal – aluminum	211
Metal – copper	400
Plastic (unfoamed) – general range	0.1–0.9
Plastic – polypropylene	0.1–0.22
Plastic – foamed polystyrene, expanded styrofoam	0.03
Plastic – epoxy resin (thermoset)	0.88

4.2.6.4 Fountain Flow

It has already been discussed how the rheology of the injection moldable material is controlled by the temperature of the melt, and the pressure that is applied to it. Polymer processing operations are also affected by the direction of flow and in injection molding specifically by the dynamics of the mold filling process, which is called fountain flow. This phenomenon is illustrated in Figure 4.8.

In order for molten material to move from the injection unit to the mold tool cavity, it is necessary to exert a sufficient force to overcome the resistance to flow. This high pressure subsequently dominates the mold filling process as the high pressure in the melt tries to equalize with a low-pressure region in the empty cavity. This pressure gradient controls the process and so in order to allow for a longer flow path it will be necessary to have a greater pressure of injection. Therefore, given these conditions in the process, wall thickness, shear rates, initial viscosity, and temperature all have an impact on the process dynamics.

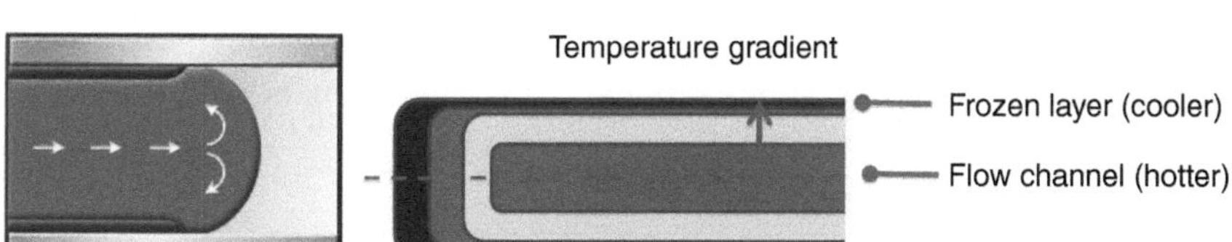

Figure 4.8 Fountain flow (left) produces a stratified temperature flow channel (right).

As the dynamics of mold filling affect the orientation of the molecules, they also affect any associated filler materials such as glass fiber in the same way as fountain flow dominates the pattern of mold filling. This is all relative to the shear flow of the melt as it first enters the mold cavity, as the melt passes from the small sized geometry of the runner system and into a larger low-pressure geometry region that is the cavity. There are also smaller elongational flow effects produced by the change in geometry. The hot molten material is pushed toward the lowest pressure regions which is the mold surface area nearest the gate, the tool surface is cooler than the melt and therefore the material begins to solidify. As further melt enters into the mold this shears the frozen layer creating a temperature gradient of hot molten core and cool solidifying mold wall as the wall area is subsequently filled by the incoming melt with the core filling last and staying hotter longest.

Because the mold surface freezes faster than the core region there are differences in morphology across the line from core to frozen layer as the frozen layer tends to be more highly orientated and sheared.

Because of the high level of orientation exerted on the melt during injection molding this can lead to mismatches in the dimensional shrinkage in the line of orientation versus the lines off center from the melt direction; this can also lead to part warping. Because this is a well-known dynamic in injection molding, designers conceive their parts to minimize these effects; however, it is not possible to always avoid it entirely.

4.2.6.5 Water Absorbing (Hygroscopic Materials)

A number of common polymer materials are water absorbing by nature, and therefore, it is necessary to ensure they are dry prior to use; dryers and example drying times will be covered in Section 4.3.6.

Using wet polymers can not only cause a visible marking at the surface but in some cases can cause a rapid deterioration of the properties themselves during processing. For this reason, there are standards very specific to allowable moisture content prior to molding (discussed later in Table 4.8). Post processing, polymers that take up moisture from the environment can also change properties relative to this level, with again a relative change in properties. With these kinds of materials it is important to consider whether you are looking at properties relating to the dry material (dry) or one that is more representative of the environment (conditioned).

4.2.6.6 Additives and Their Effects

There are already a large number of different polymer materials. If this is multiplied by all the possible formulations that can be achieved by the addition of additives, the injection molder is left with literally thousands

Table 4.8 Drying Guide for Some Indicative Materials

Product Abbreviation	Full Name	Maximum Allowable Moisture Content on Granulate (<%)	Maximum Drying Temperature (°C)/Drying Time (Hours)
ABS	Acrylonitrile-butadiene-styrene	0.20	90/3
ASA/PC	Acrylonitrile-styrene-acrylate + PC	0.1	110/3
LCP	Liquid crystal polymer	0.02	80/3
PAEK	Polyarylether ketone	0.05	200/4
PA	Polyamide	0.15	80/4
PBT	Polybutylene terephthalate	0.04	120/4
PBT + PC/ASA	PBT + PC/ASA	0.02	120/4
PC	Polycarbonate	0.02	120/4
PET	Polyethylene terephthalate	0.02	120/4
PES	Polyether sulfone	0.05	200/4
PP+ 25% Talc	Polypropylene + 25% Talc filled	0.03	120/3
POM	Polyacetal	0.15	120/4
PSU	Polysulfone	0.05	160/4
SB	Styrene-butadiene	0.02	80/3

of material formulations to choose from. There are many additives that are used everyday in the plastics industry, and therefore, it is impossible to give an exhaustive list; however, some of the most common additives and their purpose are given in Table 4.9.

For multifunctional molding as well as additives within the polymer, this can also include other materials added in-mold in spray or film form. These will be covered in Section 4.4.1.

Table 4.9 Common Plastic Additives and Their Use

Additive	Purpose
Antistatic agents	Reduce dust build up such as for products on shelves
Antislip agents	To increase grip or friction
Conductive fillers	Improved electrical and thermal conductivity properties
Coupling agents	Improved interface bonding between two materials
Fillers	Reduce polymer cost
Flame retardant	Improve flammability
Foaming agents	Produce a cellular structure for cost or rigidity
Heat stabilizers	Prevent heat degradation
Light stabilizers	Prevent light degradation
Lubricants	Improve lubrication
Pigments	Add aesthetics appearance
Plasticizers	Reduce processing temperatures
Reinforcements	Increase strength properties

Additives within the polymer can also alter the injection molding process as they can affect both the flow and heat transfer properties of the injection molding material. Therefore, even for compounds using the same base polymer (such as polypropylene), there will be a difference in the injection molding parameters of, for example, an unfilled material in comparison to a highly glass filled formulation.

4.2.7 Material Conclusions

The intrinsic material properties of the melt such as the molecular structure and therefore melt temperature and crystallization can all have an impact on the outcome of injection molding in terms of orientation and shrinkage. Additives can change the behavior of the polymer during injection molding by altering flow and heat transfer and therefore cooling rates.

The fountain flow with associated pressure, and temperature gradients mean the process dynamics are complex. When any material is already placed inside the tool surface prior to molding, the incoming molten material will shear against it as if it were the mold wall instead. The difference is that instead there is a material-dependent barrier with different thermal characteristics. A plate of steel inserted in a steel mold is unlikely to affect the process, and such processes can be used to modify the thickness of simple tool plates easily and rapidly; however, doing something similar with a sheet of plastic will affect it relative to the thickness of the sheet and the temperature gradient it induces.

As well as these material factors, there are also a range of injection molding parameters such as barrel melt temperatures, injection speeds, tool temperatures, and cooling time that need to be considered. There are also the cavity geometries and therefore the shear profile of the cavity being filled to consider, as well as material-dependent injection molding parameters. All these factors must be balanced for a successful injection molding operation. These will be described in more detail later in this chapter.

For now, we return to the placement of another material such as a sheet of plastic in the mold prior to carrying out the molding process. There are a number of issues here relating to adhesion of the sheet onto the polymer. The first consideration is the ability to surface entangle and somehow create a bond. This could be in the form of a chemical or mechanical bond, but for any interlock to be formed there must be some kind of mixing process and the two materials can stick together. That assumes of course that adhesion is required.

4.2.8 Considerations in Using Polymeric Materials as Substrates

4.2.8.1 Adhesion- or Not Adhesion

In components made up of more than one material, the properties at the interfaces become of key importance. In some designs where joint movement is required it may be necessary to use materials that will simply slip over each other; in other applications, it will be necessary to maintain integrity in the structure by ensuring there is good adhesion. Some plastic materials naturally stick to each other. For example, as long as the temperature is high enough there should be little problem heating and bonding one material to another of the same material type. An example is a polypropylene film or label in a tool with a polypropylene injection molding. However, if a nylon label was used it would not stick to the polypropylene as these two materials do not naturally adhere to each other. In multimaterial applications it is, therefore, important to know if you require a good bond or not, and to design accordingly. Compatibility tables to aid in these kinds of design decisions are available for common material types.

4.2.9 A Revisit of Thermosets

The use of thermoplastics elastomers in conjunction with thermoplastics has already been covered here; it is, however, worth a revisit of thermosets to consider differences in processing of these two distinct classes of materials.

Thermosets are injected into a hot mold as they need heat to initiate a cure cycle, and must be near cure completion before ejection. An example of this profile is shown in Figure 4.9. On the other hand, thermoplastics are heated to enable material flow and once injected they are required to

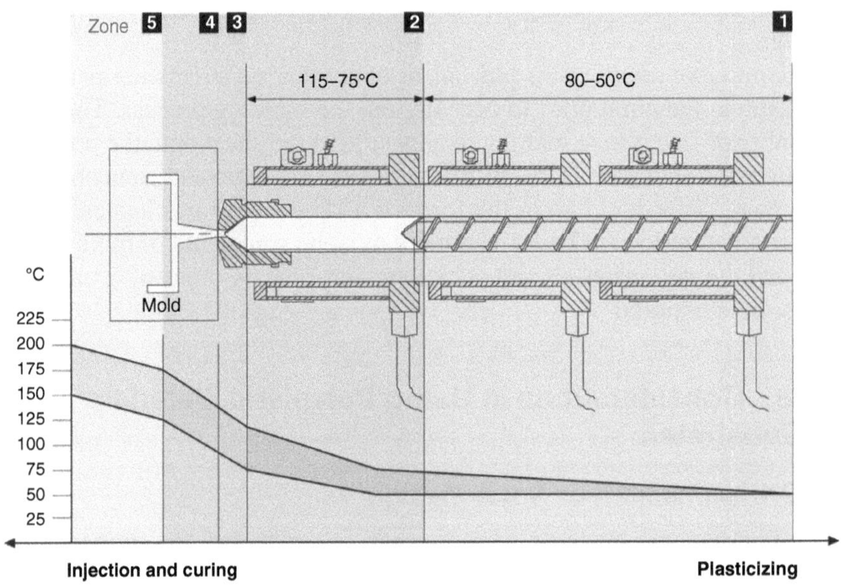

Zone 1	Preheating zone	Effect of screw geometry, rotational speed, back-pressure, and cylinder temperature control
Zone 2	Plasticizing zone	
Zone 3	Friction dependent upon injection capacity (injection pressure and injection speed) and nozzle bore	
Zone 4	Friction caused by sprue and gate cross-section	
Zone 5	Effect of mold heating and curing	

Figure 4.9 Thermoset molding temperature profile. Picture: ARBURG.

cool down as fast as possible for economic reasons, before they can be finally ejected without distortion. There is intrinsically a difference in the processing requirements. However, this does not mean that combinations cannot be processed together. There are two possible mechanisms for this to occur by, either a thermoplastic is molded and then placed in a thermoset mold or the exact opposite a thermoset is molded and cured and then placed inside a thermoplastic mold.

There are advantages and disadvantages to each method depending on the properties of the particular material involved, if for example a thermoset requires a very high temperature to cure it is likely to be highly unsuitable for combination with most thermoplastics materials. With placing a cured thermoset into a thermoplastic mold the issue becomes one of adhesion – how do you get adhesion onto something that is already cured. A solution of mechanical interlocking rather than chemical bonding may be required in this case.

One such example of this is found in the automotive industry to give parts the appearance of being made of premium carbon fiber thermoset materials. On the surface only carbon fiber can be seen to give a high quality and aesthetically pleasing finish. To produce these parts a thin cured and formed carbon fiber sheet can be placed into an empty injection molding cavity and a material such as ABS or PP can be back injected onto the carbon fiber laminate to create a hybrid thermoset/thermoplastic component. The adhesion is created either by mechanical interlock or sealing edges can be designed into the cavity to hold the materials together.

4.3 Injection Molding Machine and Process

Having discussed materials and materials factors, it is now time for a more detailed look at the injection molding process. This section begins with a deeper look at a basic machine shown in Figure 4.1 and the individual machine parts that make up a working machine.

4.3.1 Purpose

The purpose of an injection molding machine is to provide a precise and controlled dose of polymer material at a sufficient injection speed while providing sufficient clamp force on the injection molding tool to prevent the tool opening under the pressure of injection. The material needs to be homogenously mixed and molten in order to flow.

For these purposes the design of a machine can be specifically split into four parts. The **Clamp unit** which holds the mold tool. The **Injection**

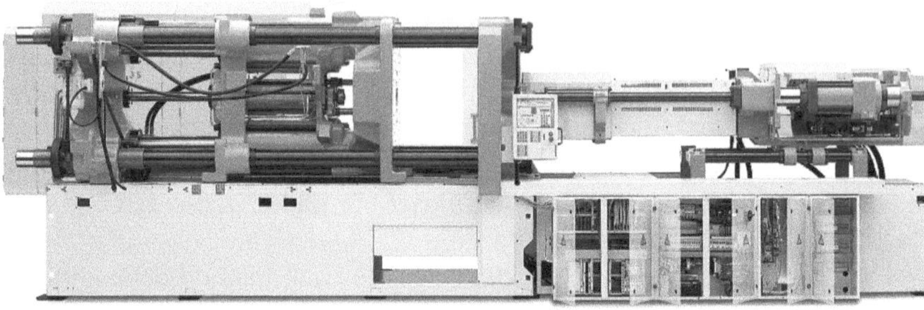

Figure 4.10 Hydraulic injection molding machine. PHOTO: ARBURG.

unit which prepares and delivers the molten material to the tool. These individual units are rated separately and along with the **machine base** with the hydraulic unit and/or electrical power supply and the **control units** and cabinets make up the basics of an injection molding machine as shown in Figure 4.10.

The machine base houses the entire injection molding system and gives the "footprint" of the size an individual injection machine requires in terms of physical space for operation.

The control unit allows both close control of the system, and the operator to programme and monitor the production operation.

There are a number of specific terms used to describe injection molding machines and processes that are included in Table 4.10.

Table 4.10 Common Injection Molding Terminology

Clamping unit	The clamp unit is required to generate enough clamping force to hold the mold tool shut during the injection molding process. See section for details.
Daylight	The dimension on fully opening the mold tool between the mold platens. Important for ejection systems.
Electric machine	A machine driven by an electric-based power system allows faster cycling and production outputs. Common in packaging industry.
Ejector unit	The system unit to eject the part from the tool after molding.
Fixed platen (stationary platen)	The side of the mold tool fixed to the machine base that does not move during the mold cycle.

Table 4.10 Common Injection Molding Terminology *(cont.)*

Flow length	The length the injected material travels in the tool cavity from the injection point.
Gate	The point at which the material travels from the sprue into the mold cavity. Many gate designs are available.
Hybrid machines	Machines combing both hydraulic and electrical drive systems.
Hydraulic clamp	Mold clamping powered by hydraulic power.
Hydraulic machine	A traditional machine driven by a hydraulic-based power system.
Injection cylinder	Part of the injection unit that houses the screw.
Injection screw	The screw from the injection unit which both doses and injects the material.
Injection unit	The injection unit is the material preparation unit and is required to take a solid polymer feedstock and deliver a controlled dose of molten material at high injection speeds into the injection mold tool. See section for details.
Microinjection machine	A type of machine with a very low shot volume. Used to produce micro sized parts.
Mold (tool)	The cavity and its housing.
Moving platen	The side of the mold tool that opens and closes.
Nozzle	Connection between cylinder and sprue bush
Parting line	The point where the mold cavity splits and opens
Projected area	The surface area of the part at the parting line that gives indications on the clamping force required to keep it shut.
Plunger unit	A nonreciprocating screw that just pressures the melt.
Reciprocating screw	A screw that turns and pushes melt forward.
Runner system (hot or cold)	A system connecting the nozzle to the cavities.
Stripper plate	A type of part ejection system.
Sprue bush	Acts as the connector for materials feed to the mold tool to the injection nozzle.
Tie bars (tie rods, tie beams)	Part of the machine base that holds the mold tool and adds to the rigidity of the base system.
Tie bar less	A machine base system that does not need tie bars.
Toggle clamp	A type of clamping system.

There is an International Standard relating to the designation of injection molding machines.

Designation 1: clamping force (kN)

Designation 2: shot capacity of injection unit = maximum injection volume (cm^3) × maximum injection pressure (bar)

1000

From this information it is, therefore, possible to immediately judge the maximum clamping force and capacity of any injection molding machine.

The various parts of the injection molding machine will now be described in more detail, starting with the injection unit.

4.3.2 Injection Unit

The injection unit serves two functions: it is required to feed, prepare, and dose material consistently and accurately and also to then inject the set volume of material at high pressure into the mold tool.

It consists of several distinct sections: the hopper and feed throat section, the screw section, and the injection nozzle and is shown in Figure 4.11. There are a number of injection molding parameters that relate to the action of the injection unit. These include setting the melt temperature, setting the correct shot volume, setting the injection speed, and the mixing

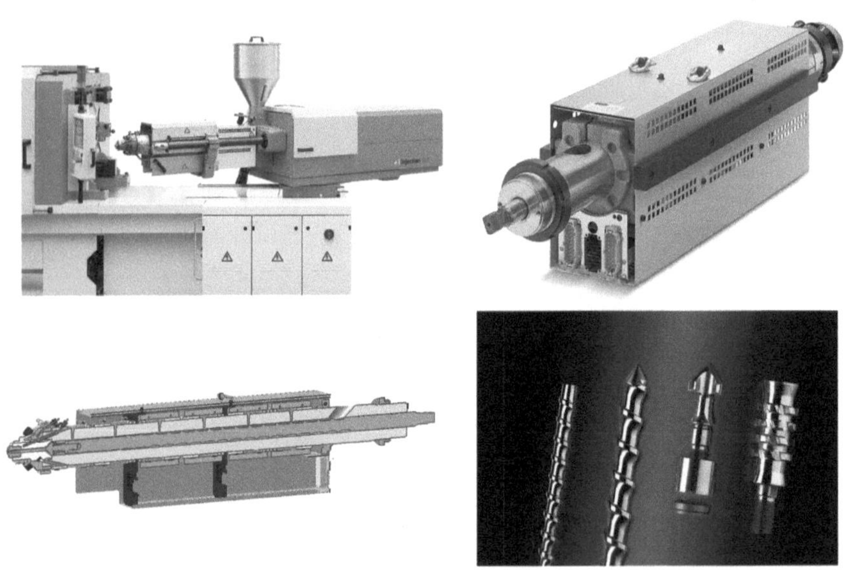

Figure 4.11 A complete injection unit (top left), the nozzle and heating covers (top right), nozzle and casing removed to show screw (bottom left), screw designs (bottom right). PHOTOS: ARBURG.

speed of the screw that is controlled by the screw back speed. This is called the back pressure. The injection unit carriage can also be moved to either sit directly on the sprue during injection, or to be retracted during dosing and purging if required.

4.3.3 Clamp Unit

The clamping unit is housed in the machine base as shown in Figure 4.12, and its purpose is to hold the tool rigid and shut during moulding.

Injection molding is a very high-pressure process and therefore very high clamping forces are required compared to other thermoplastic production processes. The level of clamping required is related to the surface projection of the molding; larger surface areas require greater clamping force. Therefore for the production of larger parts, larger clamping forces and therefore larger clamping units to generate these forces are required.

Setting sufficient clamp force, controlling the opening and closing speeds, and setting sufficient daylight for parts to be ejected are all functions related to the clamping unit and the particular tool being used. The speed of movement of the clamp unit can affect the overall production volumes and efficiency. Greater speed of movement can be achieved with electric clamping units and therefore for very high volume and fast production an electric clamp unit may be advantageous in some circumstances. However, the choice between hydraulic or electric machines is often made due to a variety of other reasons such as energy efficiency, investment policy, or even political considerations.

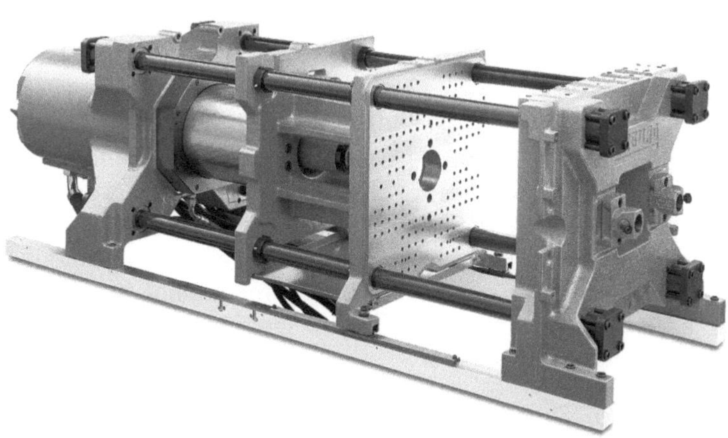

Figure 4.12 Electric clamp unit (left) and hydraulic clamp unit (right). PHOTOS: ARBURG.

4.3.4 Tooling

Once the capital costs have been met for machine and ancillaries, the single most important investment for any individual injection molding job is the design, build, and purchase of each injection mold tool. Every new component needs a dedicated precision tool, the cost of these tools varies enormously based on their size and complexity but can reach costs in excess of a million UK pounds for large complex tools. Not surprisingly, this cost is a very major factor in the profitability of any single commercial injection molding job.

Due to the high clamping forces, shears, and complexity of injection molding it would be impossible to cover all aspects related to tooling and there are a number of more specialized publications solely related to this discipline. There is an industry in itself related to the technical design and simulation of the molding process and sometimes commercial molders have this function of specialist tool design and engineering shop in-house for this purpose. There are also many companies that solely design and supply this tooling to the manufacturing sector. Therefore, there will only be a very basic introduction to this subject provided here.

For this purpose injection mold tooling can be considered consisting of two or more steel parts that when closed form the cavity shape and injection flow path.

It must provide the following functions:

- Form the cavity into which the molten material is injected
- Act as an effective heat exchanger to remove heat from the surface of the cavity
- Have sufficient rigidity to seal the mold halves under injection pressure

Given these factors it can be seen the quality of the final component produced is highly dependent on tool design and therefore tooling is a vitally important process for any new product introduction. The principles of how tooling works, gating, ejection systems, and methods to maintain temperature control that all go to make up a functioning mold tool are all introduced here. Again, those wishing to learn more are directed to the recommended reading.

4.3.4.1 Two-Plate Mold

The simplest and most standard design consists of two sides, one is just a flat surface (male) and is called the stationary plate. This is as it does not move during the injection molding process. This plate contains a flow path from the injection unit to the cavity for the molten material and this is called the sprue. The other plate contains the cavity shape (female) and moves to open and close the mold to allow part ejection. It also contains the ejector system. The interface where the two plates meet is called the parting line. This is shown in Figure 4.13 (top left).

4.3.4.2 Stripper Mold

The addition of a stripper plate distinguishes this design from the two-plate mold that uses individual ejector pins to eject the rigid component from the cavity. In some designs using pins can distort and stress the component in the pin regions during ejection, a stripper plate allows for a much higher surface area and reduces the chance of introducing stresses.

A comparison of a stripper mold design (top right in Figure 4.13) with a standard two-plate mold (top left on Figure 4.13) can be seen by referring to the figures.

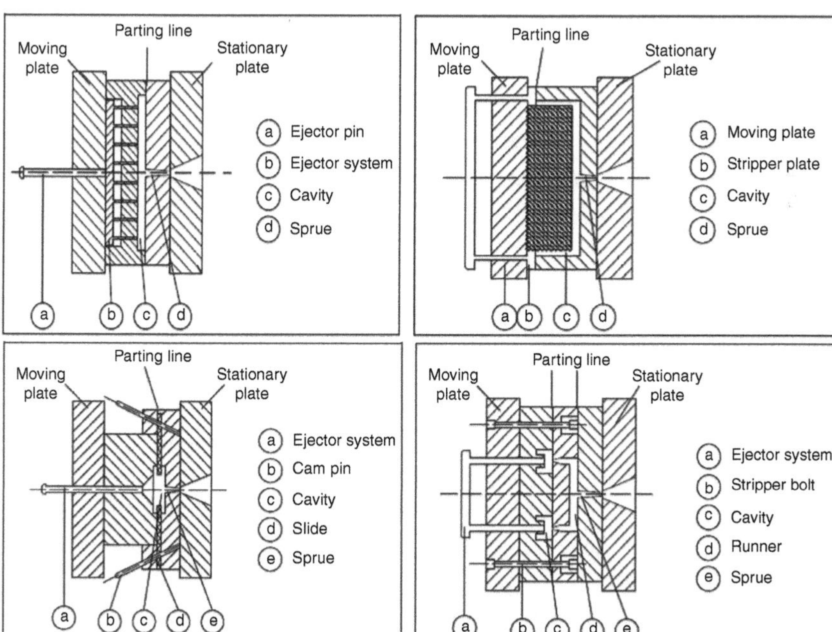

Figure 4.13 Mold tool variations top left (two plate), with stripper plate (top right), slide mold (bottom left), three-plate mold (bottom right).

4.3.4.3 Slide Mold

A slide mold is also still a two-plate mold; in this case, it adds the complexity of sliding sections for additional movement in the lateral direction. This type of movement allows for designs with undercuts to be produced or, for example, components with external threads.

For a comparison of slide mold design (bottom left in Figure 4.13) with a standard two-plate mold (top left in Figure 4.13) refer to the figures.

4.3.4.4 Three-Plate Mold

If more than one component is being produced at a time and the tool has multiple cavities it is necessary to split the sprue into multiple flow streams; to do this, as the name of this tool suggests, a three-plate mold is used. This plate usually sits next to the sprue on one side and houses the additional feed system for the flow path that is required; this is called a runner system that feeds each individual cavity from the original sprue on its other side. As with other deigns the stationary part again carries the cavity form as a female part design.

For a comparison of three-plate mold design, see Figure 4.13.

Opening of a three-plate mold requires separation of the component (moving plate) and the extra third plate by the introduction of an extra parting line. The tool opens to eject both the sprue and runner on one side and the component on the other. In effect as if there were two two-plate molds in operation here. As it is necessary to ensure ejection of both component and waste (sprue and runner), these systems tend to run with a slight delay from one side to the other making production slower than with a two-plate mold; however, the advantages are multiple components that can be produced during one cycle.

There is an issue of scale that must also be considered with multiple cavities as by default a multiple cavity will need to be a larger injection molding tool than a single cavity producing just one component. It will, therefore, probably need a larger injection molding machine for production with a greater clamping force; the tool will also be heavier due to the greater volume of tool steel needed. As well the runner systems must be balanced to ensure all the multiple cavities produced are of identical weight and size. There is also higher production wastage per injection cycle as both the sprue and runner are discarded. There is a tradeoff between the number of cavities, the cost of the tooling and machine capacity to be considered here. Additional items may also be involved here such as external screw pickers to aid ejection as shown in Figure 4.14; there may also be robots coming in and out of more complex operations that will be described in later sections. An overview of the factors that need to be

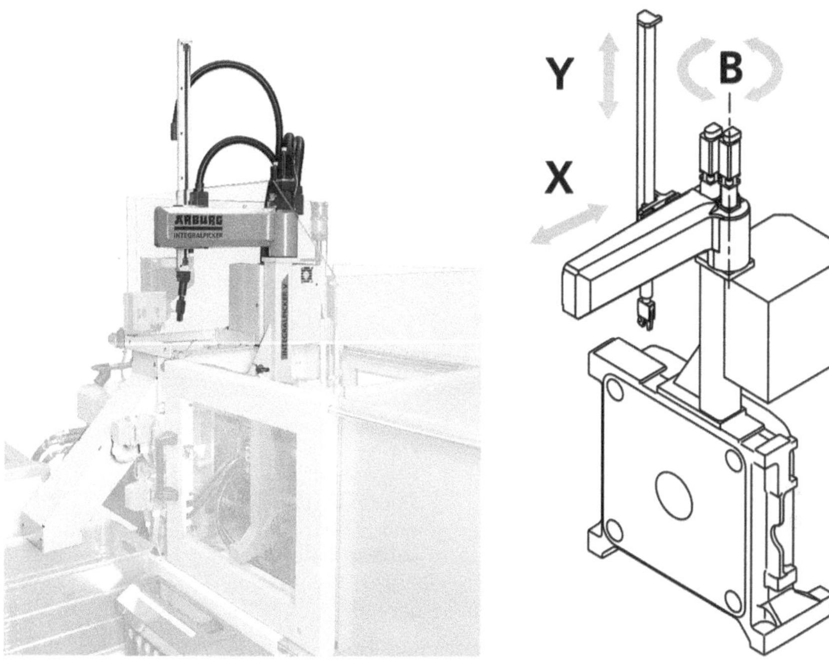

Figure 4.14 Robot arm for sprue picking or part removal. PHOTO: ARBURG.

considered in tool design and that have been described in this section is illustrated in Figure 4.15.

For multicomponent molding, tooling systems can also consist of more than one injection molding tool. Rotating platens can be used on the moving side to rotate individual tools between cycles to build up more complex components. There may also be additional robotics integrated with the system to move components from mold to mold or place inserts into the tool before molding. This will be discussed in more depth in later sections.

4.3.5 Mold Cooling

With thermoplastics the main purpose of the mold system is to minimize both the cycle time and thermal differences in mold part cooling. Mold cooling is, therefore, essential for both cost saving and quality control. Uniform cooling improves product quality by preventing differential shrinkage, high residual stress, and mold release problems. Therefore in designing an injection mold tool, the size and layout of the cooling channel are of utmost importance.

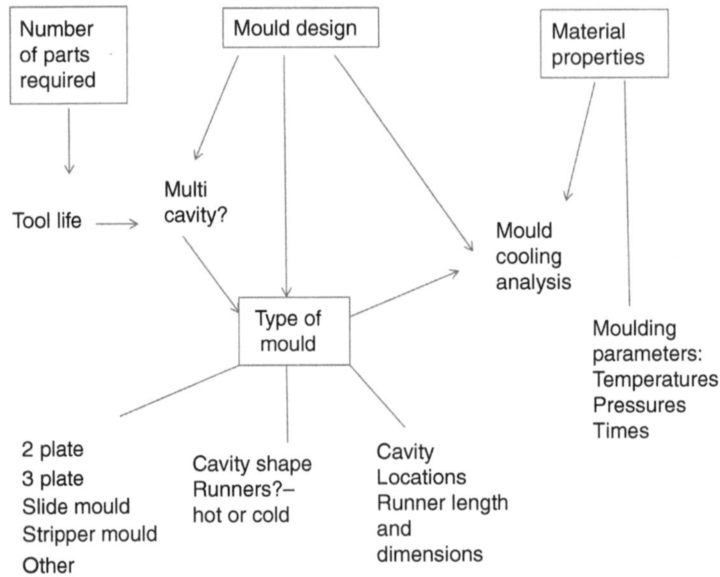

Figure 4.15 Overview of tooling considerations.

The cooling time between part injection and part ejection accounts for at least 60% of the injection molding cycle. Therefore, this cooling time plays a vital role in the efficiency, quality control, and productivity of any component–tooling combination. While reducing the cooling time can benefit the economic viability of production, it is necessary to retain a uniform cooling process for component quality as cooling time can impact residual stresses, differential shrinkage, mold release, and also the ejection system. The importance of the cooling stage has, therefore, led to a lot of research to attempt to speed up the heat transfer processes. For multimaterial molding, this time becomes even more vital since often by placing one material on top of another, a further thermal barrier is created between the injected molten material (low-conductivity polymer material) and highly conductive tool steel.

Therefore, it is worth spending time considering the cooling design, especially in respect to the implications for multifunctional material molding.

As stated an efficient cooling circuit is vital not only to reduce residual stresses, but also to maintain part to part dimensional accuracy and stability. Temperature variations in the cooling cycle are, therefore, to be avoided wherever possible.

A typical cooling system is shown in Figure 4.16 with the mold tool acting as the heat exchanger, with the heat from the molten polymer being

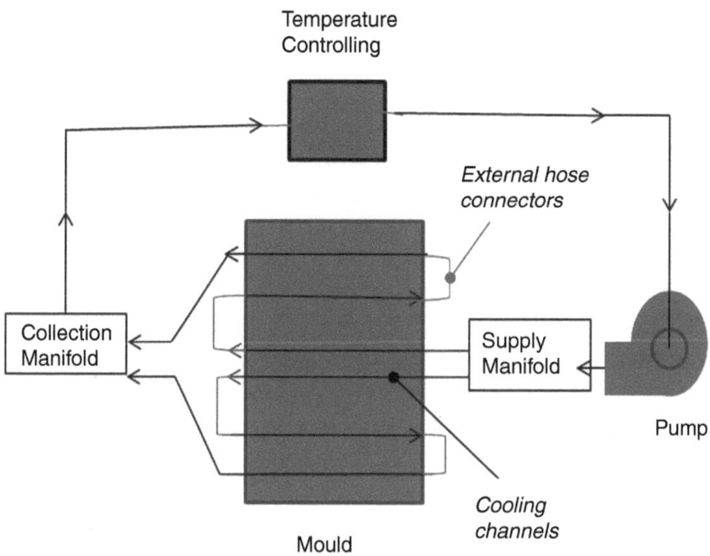

Figure 4.16 Typical cooling system for injection molding tool.

carried away by the coolant circulating in the cooling channels and back to the heat control unit.

The easiest circulation systems use water; this use is limited by the boiling point and freezing point of water, if higher temperatures are needed, oil-based systems can operate up to 350°C. For chilled temperatures below 25°C, it is necessary to use water mixed with antifreeze (glycol).

Other more advanced methods for heating and cooling injection molding include the use of induction heating, high pressure water systems, and partial tool heating; cooling requirements and therefore cooling times should, wherever possible, always be optimized using the best solutions available.

Injection mold tool designers need to bear in mind the importance of cooling the part, especially in terms of wall thickness, as doubling the wall thickness of a component can have a quadruple effect on the cooling time due to the low thermal conductivity of most polymer materials. Therefore, as a rule of thumb, as the thickness increases the cooling time can be considered to increase by a factor of $1{:}1^2$. (This would not apply to specialized conducting polymers as described in Chapter 2).

Consider the following properties of an injection molding material:

$$\text{Thermal diffusivity} = \frac{(\text{Thermal conductivity})}{(\text{Density})(\text{Specific heat})} \tag{4.1}$$

Cooling time is proportional to the square of the heaviest part wall thickness divided by this thermal diffusivity value, this can be used to provide a general guide to the cooling time.

The cooling channels are drilled into the mold tool during manufacture and are fitted into both the moving and stationary plates in order that the cooling on either side of the parting line is equal. Mismatched temperatures on tool faces are to be avoided, not only for the impact on the injection molding component but also as this can cause one side of the tool to expand relative to the other, thereby preventing smooth operation. If this temperature difference is excessive, it can prevent the mold closure process entirely. Ideally, there should no more than 10°C difference if possible, the less difference the better. The cooling channels are best placed as near to the cavity as possible and are therefore ideally located in the same block of metal as the cavity itself. However, this should not be so close so as to affect the integrity of the mold.

To more accurately specify the design of the cooling channels, it is necessary to consider factors such as:

- Where they will be located (and depth and width from mold surface)
- Size of cooling channels (width and depth)
- Type of cooling channels
- Layout and connection
- Length of circuit
- Flow rate of coolant

As the cooling channels move away from the cavity there is a reduction in the heat transfer efficiency.

As the distance between each cooling channel increases, the nonuniformity of cooling will also increase.

An example of a cooling unit is shown in Figure 4.16; standardized equipment is readily available using both water or oil if higher temperature usage is needed. Both liquids can be used as coolants and various commercial units contain all the equipment shown in Table 4.11 allowing fast control. Standardized sizing and fittings by the industry allow easy connection with hoses by quick fit connectors. As previously mentioned, developments in decreasing the cooling time and therefore overall cycle time have meant there are also heating and cooling systems using hot gases or steam within the cooling channels as alternative coolants. These pressurized systems can often be rapidly switched over to operate "cooling" mode or "heating" mode as required enabling much closer and more accurate control of the cooling cycle. There are also noncontact methods using

Table 4.11 Mold Temperature Settings: Some Indicative High and Low Control Values for Plastic Materials

Short Designation	Long Name	Demolding Temperature Ranges (°C)	Coolant
ABS	Acrylonitrile-butadiene-styrene	35–90	Water
FEP	Tetrafluoroethylene hexafluropropylene	160–220	Oil
LDPE	Low-density polyethylene	30–65	Water
HDPE	High-density polyethyelene	40–75	Water
PC	Polycarbonate	60–130	Oil
PP	Polypropylene	45–80	Water
PA66	Nylon 66	75–150	Oil
TPE-E	Thermoplastic elastomer (type E ethylene),	25–65	Water/ethylene Glycol*

** Ethylene glycol (antifreeze) is commonly mixed with all water systems to help prevent bacteria growth.*

resistive heating such as electrical resistance or magnetism to heat and cool the tool, a process termed induction heating. Infrared radiation has also been investigated as a heating method.

An alternative, but highly effective, way to get around long cooling times is to simply move or rotate the tool away while it cools allowing the opportunity to shift another mold in and mold another component. Such systems are commonly used for multicomponent molding and will be introduced later in the chapter.

4.3.6 Ancillary Equipment

4.3.6.1 Dryers

Hygroscopic materials need to be predried prior to injection molding. The market contains a number of options for drying that also include built-in delivery systems to feed the material to the hopper as required.

Whether a material requires drying before molding may not only depend on the sensitivity of the material to moisture but also to the previous storage conditions of the material. Material stored outside may need predrying regardless of whether it is hydroscopic in nature, simply to keep moisture levels low in the injection molding system. This not only needs to be consisted for the injection molding material itself, but also for any other materials (e.g., inserts) introduced into the system elsewhere.

There are a number of drying solutions such as desiccant dryers that are considered the most reliable and economical method of predrying methods and are not affected by local atmospheric conditions. Circulating air ovens can be also used, but with caution and with consideration to local humidity levels. Vacuum dryers are efficient but expensive as long heating times are needed when compared to desiccant dryers.

Injection molding machines can also be fitted themselves with degassing units, but this is not as efficient as predrying, especially for sensitive hydrolytic materials that may degrade before they can begin to degas. However, this method can be very useful when dealing with less sensitive materials or recyclate materials.

A further consideration for drying is the conveying systems used to transfer material from storage and drying units to the machine hopper. As some resins can reabsorb moisture very quickly (e.g., PET may reabsorb excess moisture after drying, after only 10 minutes in the atmosphere), the times the material can dwell in either the hopper or the conveying system without protection from moisture ingress also need to be considered.

4.3.6.2 Introducing Robots

Robot use in the injection molding industry is one of the current growth areas as more complex operations are integrated into the injection molding process.

Robots can be used to control insertion and removal of components as part of an automated production process. For processes such as insert molding, very accurate placement and part registration is required to achieve the level of precision necessary, which is often a combination of advanced tool design to hold the insert in place, as well as the programming of the robots for accurate and repeatable placement and removal processes. Further multifunctional aspects of robots are considered in Section 4.5.

4.3.7 The Thermoplastic Injection Molding Process Cycle

Injection molding is a continuous but cyclic process that is set to run automatically to produce literally thousands of identical parts in any production run. Once set up and running correctly, including an automatic parts handling process for molded items, a machine should operate completely automatically with little need for human interaction other than perhaps the quality control function.

4.3.7.1 The Cycle

The production of a single component from start to finish (part ejection) is called an injection molding cycle. It is, therefore, the setting of this single cycle of operation that is of importance and allows modern injection machines to repeat this cycle with great precision thousands and thousands of times over.

For thermoplastic materials the process stages work as follows:

1. The injection mold tool is closed and clamped shut.
2. A controlled dose of material is injected at high pressure into the cavity by the injection unit.
3. Holding pressure (to account for shrinkage) is applied if needed.
4. Cooling time is applied to allow the component to become rigid enough to eject without distortion.
5. The injection unit moves back (optional process).
6. The injection unit is dosed for the next process cycle.
7. The tool is opened.
8. The component is ejected and the cycle begins again with stage 1.

The total time for an injection molding cycle is both tool and material dependent and therefore highly variable; however, an indicative injection molding cycle showing the various relative timings of each stage can be seen in Figure 4.17. It can be seen that the cooling of the molding before

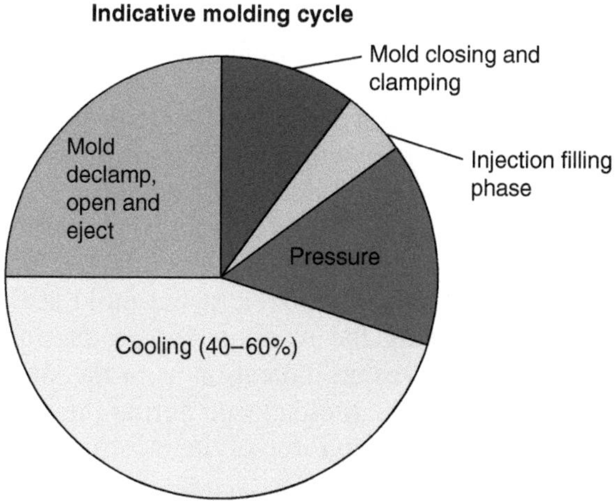

Figure 4.17 Component times of a typical injection molding cycle.

part ejection is by far the biggest portion of the overall cycle and therefore has economic implications in commercial use.

4.3.7.2 Variable Parameters in the Injection Molding Process

A single injection molding cycle, once set correctly, can be repeated and operated with great precision thousands and thousands of times with correct machine settings. The consistency of injection moldings, and thus the production of identical parts, therefore relies on the injection molding machine parameters being optimized and set correctly.

As discussed in the material sections, the high viscosity of polymer feedstocks requires high pressures to build up in the system before injection of the material at the correct injection speeds. Speed, along with temperature and therefore the shear rates, impacts the viscosity of the melt. The extreme forces of injection molding also mean that a high clamp force is needed on the tool simply to prevent the injected material from forcing the tool open during mold filling.

A cycle involves some of the following operations to be set by the setter as shown in Table 4.12. Incorrect settings to parameters such as shot weight, cycle time, melt temperature, mold temperature, cooling temperature, and injection speed can all cause inconsistent production parts to be produced in terms of varying component weights, incomplete shots, or overfilling. Generally, the clamp force is set whenever a new tool is installed and would not be changed again if operating correctly.

4.3.7.2.1 Wall Thickness and Projected Area

At this point it is also worth touching on two important design considerations for efficient production; these are the wall thicknesses of the component and the surface area at the molding tool split line.

It is important to consider wall thickness in injection molding design; this is because as the wall thicknesses increase, the injection pressures required also increase. Therefore, the flow path/wall thickness ratio and the viscosity of the injected materials are important design factors for efficiency. The surface area of the mold at the parting line is also important for sizing the machine for production. The surface area at the parting line gives an indication as to the size of the clamping force required to keep the mold closed during the injection molding process and therefore the clamp force of the injection molding machine required.

Table 4.12 Injection Molding Parameters Related to Operation of Injection Unit (light grey) and Machine Base Control (dark grey)

Temperature (°C)		Injection Unit Barrel Temperatures: Throat, Zones, and Nozzle				Mold Temperature		
Pressures (Bar)	Injection Pressure Max.	Back Pressure		Holding Pressure		Clamp Force	Ejection Pressure	
Times (s)	Plasticizing time	Screw back time	Injection time	Holding time	Injection unit delay	Tool opening and closing time	Cooling time	Ejection time
Weight (cm^3) or distance (cm)	Injection shot size	Screw travel	Holding set point	Melt cushion		Tool daylight distance		Ejector travel distance

4.3.7.3 *Product Design*

The wide scope of conventional design for plastic materials is best considered with reference to the designers themselves. Chris Lefteri published "*The Plastics Handbook*"in 2008, which covers a broad and imaginative scope of the best uses of plastics materials in design.

When designing for injection molding there are a wide number of considerations for aesthetic, function, manufacturing constraints, and the interrelationships that exist between the design, materials, and injection molding process not least including the injection molding tool itself. Figure 4.15 and Table 4.12 give an introduction of just some of the tooling and material-specific considerations within any new injection molding project.

4.3.8 Process Variations

4.3.8.1 *Microinjection Molding*

A further variation on injection molding is microinjection molding. This is used to produce very small micro-sized components such as tiny gears, cogs, and other small structures. Unlike injection molding, micromachines utilize a plunger system rather than a reciprocating screw and generally produce parts of 1g or less. Microinjection molding is newer on the market and therefore less developed than the convectional injection molding market; however, many of the processes discussed here will be applicable to some extent in the micromolding arena. Specialized microinjection molding machines are available, is it also possible to modify larger machines by piggy backing a smaller microinjection injection unit onto a larger machine. An example of this is shown in Figure 4.18.

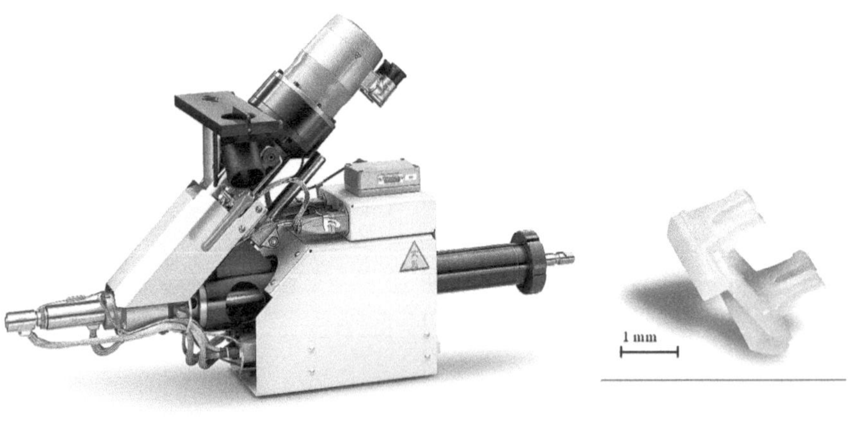

Figure 4.18 Microinjection unto for a piggy back system (left), example microinjection part (right). PHOTOS: ARBURG.

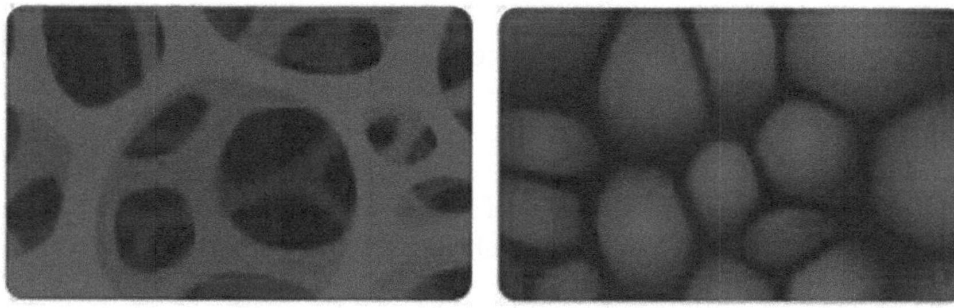

Figure 4.19 Cellular structures: open (left), closed (right).

4.3.8.2 Foam Molding

A common processing technique to lightweight and add rigidity to the structure of an injection molded component is foam molding. Additives called foaming agents can be mixed into the melt to produce foamed structures, that is air spaces within the plastic structure. The gas-producing additive can be mixed with the plastic pellets in the hopper or gas can be inserted straight into the molten melt using specialized equipment. Dependent upon the action of the additive, and/or the mixing in the melt an open or closed cellular structure can be produced as shown in Figure 4.19. Using a method of direct gas injection into the melt, which needs a specially designed screw unit, allows close control of the microcellular foam structure that is produced. This technique is termed the MuCell process. This type of foaming process should not be confused with assisted molding techniques such as gas or water injection that produce hollowed-out components by a multimaterial molding technique and will be discussed later.

One disadvantage of foamed structures is the quality of the surface finish as it can be difficult to achieve a smooth surface with this technique. However, it can be used in conjunction with other processes to create multimaterial or multifunctional products with improved surface qualities.

4.3.8.3 Assisted Molding (Gas and Water)

If instead of mixing a gas with plastic to produce a foamed structure, it is kept separate to form a stratified and laminar layer of polymer/gas/polymer; a hollowed-out structure instead can be created. This is essentially the purpose of assisted molding. Whether considering using a high pressure gas or high pressure water injection, the main feature of assisted molding is the use of a nonpolymeric component as the second injected component. The second component may be introduced either through the nozzle or more usually by a separate injection process with a mold module and a tool-based injection pin which is inserted into the mold. This allows the gases to be controlled and pumped directly into the mold tool at preset parameters that coincide with the injection cycle to produce optimized parts; therefore, these systems integrate fully to the machines to control and monitor the molding cycle.

The injected nonpolymer material is also removed in the standard cycle by design, and therefore does not make up the final component but to core out the component for cost and weight savings.

These processes bare close relation to coinjection processes that follows this section, but eliminating the core polymer material for a much lower rheology and higher flow properties of a gas or water solution.

In terms of multifunctionality, these processes can core out and lighten moldings if needed and can also be used with other advanced processes to build in multifunction.

4.3.8.4 Coinjection Molding

This can be used to produce sandwich structures.

Again, this process uses two stratified materials to produce skin/core/skin structures, known as sandwich structures. It essentially combines two different material feeds and therefore two injection units are combined by a common nozzle unit. There is some general variability within the process in that it can be used to inject both simultaneously (skin and core together) and sequentially (one and then the other). This processing method is also sometimes known as 2K (two-component) injection molding and 3K (three components) has also been realized. In each case this is where multiple injection units are combined within the same injection nozzle to produce stratified layers of material. Therefore, multiple injections are controlled within the injection units as opposed to multiple injection points or combinations in the tool. This requires some kind of dedicated coinjection nozzle that combines the flow of both (or more) injection units.

4.4 Toward Multifunctionality – in Mold Techniques and Multishot Techniques

Having looked at some basic and more advanced processing methods, it is now time to consider how these can be developed to integrate greater functionality into injection moldings.

This section will be split into two parts:

Processes that inject more than one material (more than one injection unit).

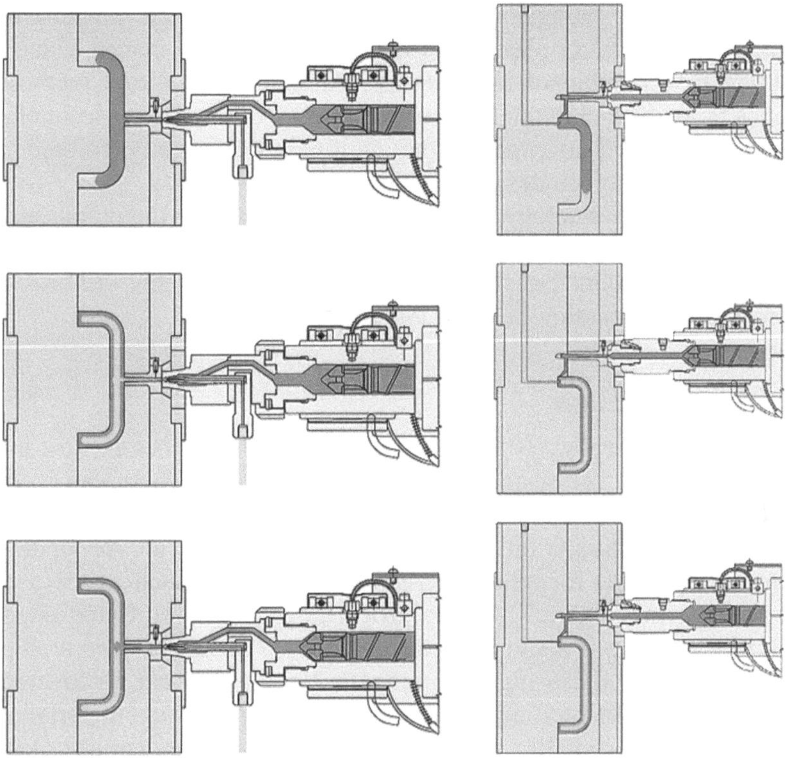

Figure 4.20 Through a nozzle (injection unit) process (left), separate injection points into tool (right).

An example of this has been illustrated by the process coinjection molding.

Processes that combine more than one material in the tool (tool controlled).

An example of this is assisted molding where gas is injected directly into the tool via an injection pin embedded in the mold cavity.

Further example of the distinction is shown in Figure 4.20.

4.4.1 Processes that Combine More than One Material in the Tool

Over molding and insert molding are both general terms that can be used where there is a molding process over a previously manufactured component. This premade component can be metal, ceramic, plastic, or any other material; the term is not material dependent or shape dependent.

Commonly seen examples include over-molding plastic handles onto metal scissor blades or plastic handles onto screw driver blades. The over-molding of a number of multicolored plastics onto inserted toothbrush bristles is another commonly seen example of over-molding technology. This can be used to produce the multitude of brightly colored and imaginative toothbrush designs commonly available today.

This kind of technology can also be used to include the preplacement in the tool of film inserts and labels, and the subsequent back injection of materials to consolidate them into the final component. This will be where the following discussion of insert molding begins.

4.4.1.1 Decoration

There are several ways in which injection moldings can be post treated and decorated for aesthetic or labeling purposes. These can include painting, plating, labels, screen printing, hot stamping, pad printing, decals, and dye sublimation. However, as the industry has responded to customer demands for new materials and processes such as the use of hybrid materials and new surface properties, it has been found integration of decoration is also possible to produce a complete decorated part and remove postprocessing requirements. New legislation, as covered in Section 4.6, is also fuelling a trend for reduced energy and material consumption and end-of-life recyclability and/or biodegradability. Again, including a finishing process in-mold can be beneficial in meeting these targets. These kinds of developments have formed the basis for a move to multifunctionality as attention has turned from simply cosmetic to more functional activity. It is, therefore, worth considering the developmental steps to explore all possibilities inherent within a basic injection molding operation.

It has been shown that in-mold technology enables various processing steps to be combined into one within the mold to achieve added functionality, improved aesthetics, or surface properties. Common techniques such as in-mold decoration and in-mold labeling can be designed to improve the aesthetic appearance as well as durability of the created plastic products. The cost savings can also be considerable due to the removal of postprocessing operations and savings of up to 40% have been reported (Lentz, 2010) as well as reductions in the total process carbon emissions. Therefore, in many (but not all) cases this can be a cost-effective process for combining manufacturing with decoration of plastic parts. A wide range of decoration effects can be achieved with multiple texture and durable surfaces. Therefore, this technology has been applied across sector and to a number of products and has shown to be particularly successful technique for interior components for automotive, consumer electronics goods, and domestic appliances.

Not surprisingly, developments are fuelled by the introduction of new products by the most fast market moving industries such as the food packaging industry and mobile phone-related industries where relatively low lifetimes of products and product turnover naturally fuel faster development in a competitive market place.

In terms of in-mold decoration there are a number of common terminologies used to describe what essentially the same process route. This is one of preinsertion of an insert of some kind into a mold tool before it is integrated during the injection molding cycle to enhance the appearance of the final product.

Generally but not always they are precluded by the word "in-mold" such

- In-mold decoration (IMD)
- In-mold foiling (IMF)
- In-mold labeling (IML)
- In-mold textiles (IMT)
- In-mold veneer (IMV)

Film insert molding (FIM) is also a term that is used. The more developed term of in-mold electronics (IME) is spin off from such techniques and will be covered in the later sections, which now continues with an overview of these technologies that are supported by most of the major molding manufacturers. It will continue with a look toward the future and some less common applications relating to the integration of electronic components, such as lighting, identification labels, sensors, and batteries. This review provides a brief overview of the technique to highlight the opportunities that this technology can offer.

As the name suggests film insert-type technologies refer to any method where an overlay or coated film becomes part of the molded product (Love and Goodship, 2002). This generic term can be used to cover a wider range of potential uses.

4.4.1.2 In-Mold Labeling (IML)

To go back to beginning of in-mold technology it is necessary to consider the labeling of an ice cream tub. It was found that adhesive was not needed when inserting paper labels into the injection mold, removing the need for an adhesive and for postlabeling processes.

From this technique further improvements were developed with the move to using polypropylene (PP) labels in PP parts. This produced a mono-material product that was and still is therefore easier to recycle at end of life. In early development, the films were precut before being inserted into the tool; further development therefore also removed this stage by taking the labels directly from a roll of film running alongside the machine using a reel-feed system. In this way, it can be seen how continuous development of technology has led to an entirely new process eventually eliminating the need for precutting. Process integration also improved the overall recyclability of the product and made cost savings through reduced manpower requirements. It is, therefore, easy to see the economic benefits accrued. Further developments have been to use multilayer film and labels to create even greater design possibilities.

It is possible to create smooth clean edges, and as the (factory environment) process can be easily kept clean or hygienic as needed, it has also gained proven applications in more demanding areas such as food, pharmaceutical, and medical industries. If suitable high-performance materials are used within these labels they can also resist cleaning and chemical and/or environmental exposure. They can also offer some protection to manufacturers from counterfeiting while providing eye-catching designs and graphics.

While IML does require a costly investment in machinery, this can be repaid when considered with other labeling processes in high cost/volume scenarios. The downside of the process in terms of injection molding processing is a longer molding cycle per part is required due to this labeling occurring at the molding stage.

IML can be used to integrate a wide variety of pattern or graphics onto the molding surface. The IML take the form of a film, which is usually prepatterned by screen printing. Preforming of the film prior to molding is dependent on the curvature and complexity of the final component. For simple surfaces preforming is not usually required but for more complex surfaces it is necessary to avoid film wrinkling or tearing.

After the film is preformed (if necessary), cut, and placed into the mold tool, it is held in the mold tool by design features built into the mold (i.e., vacuum, pins, or electrostatics), so it does not move when molten material is injected into the mold. IML are most commonly placed by robots to ensure alignment and registration is as accurate and repeatable as possible. The final product removes the need for postdecoration processes and associated costs while providing an attractive, durable, and customizable finish for the component.

A further more recent cost-saving development is inline printing (see Figure 4.21). An example of this is the InkBOT process developed by FBT, Germany. Here, digital printing and robotics combine to produce a flexible print on demand machine allowing even greater flexibility (FBT, 2015).

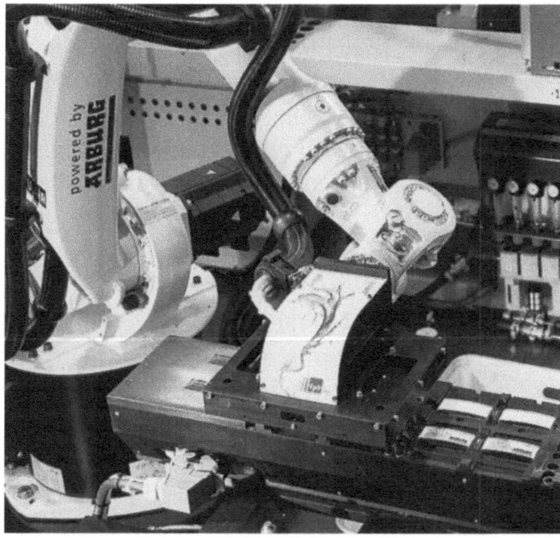

Figure 4.21 In-line printing.

This produces as even faster inline decoration process and removes the need for procurement of preprinted films and any prior treatments such as cleaning and cutting. Enabling extremely fast start-up and easy design changeovers, highly customized production can be carried out. This process can also cope with some simple geometrical design features such as concave or convex components. The printing is noncontact and produces high-resolution borderless color.

4.4.1.3 In-Mold Paint Films

In-mold films that mimic a painted surface have been used since the 1970s and have allowed the production of interior plastic components with a wide range of colors and textured finishes such as brushed metal or wood grain. They are generally of a multilayer construction and have been developed to be durable enough for exterior use, while still providing a suitable high-quality surface such as high gloss if required.

In-mold paint films have a variety of applications but have gained a major foothold in the automotive industry where they have been applied to a number of large 3D parts such as entire instrument panels, as well as bumpers and the roof module of the 2003 Smart car.

The major driver is the elimination of the energy-consuming automotive paint shop but other associated benefits can include weight reduction and increased production efficiency.

An alternative to incorporating an entire film is to just transfer a decoration film. In these cases carrier films are used with the pattern transferred while the carrier film is removed after molding. The IMD process is the same with the film still placed and held in the mold prior to injection. This type of IMD decoration is not suited to components where durability is required and can wear off; however, again complex shapes are possible (Love and Goodship, 2002).

4.4.1.4 In-Mold Textiles (IMT)

The high injection pressures of the injection molding process give it the inherent ability to penetrate deep into the fibers of many textile materials and therefore create a strong mechanical bond. It is possible to use a range of textiles in-mold to decorate injection molded components. Leather effect fabrics and carpet like fabrics, generally backed by layers of foam, can all be introduced. These are commonly used to produce interior trim for automotive applications but again can potentially be integrated into more multifunctional designs.

4.4.1.5 In-Mold Veneers (IMV)

The development and use of real wood veneers in injection molding was championed by Jaguar Land Rover in the automotive industry. Traditionally, real wood was used as a marker for quality and found only in premium quality, high-end automotive makes, and models as the use of wood veneering was resource intensive, time consuming, and costly. An alternative in-mold manufacturing technique was therefore highly viable. Makenji et al. (2006) developed an alternative in-mold production method that is now used in production, which produced considerable time and cost reductions. An example component is shown in Figure 4.22.

4.4.1.6 In-Mold Electronics (IME)

As electronic technology has rapidly developed there has been an associated explosion in the world's number of energy-using devices and products (Owen, 2011). There are a number of things driving these developments; reduction of production costs, material security and

Figure 4.22 In-mold Veneer sample.

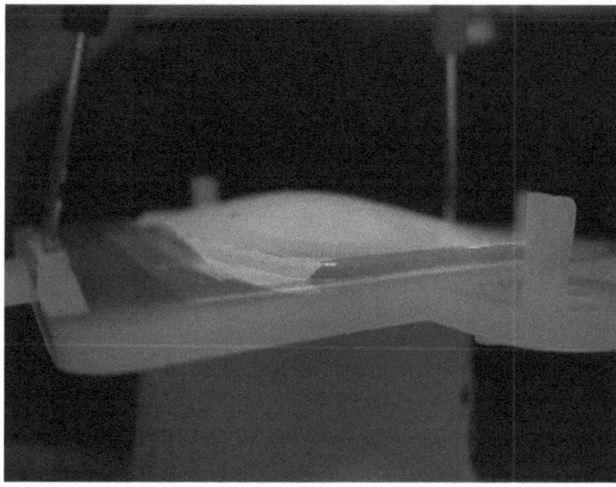

Figure 4.23 Functional EL device.

supply, environmental legislation, energy efficiency and technological breakthroughs in printed electronics, as well as the subsequent fast introduction of new products up taking the new technology. It is, therefore, not surprising that the introduction of in-mold technology has also been considered. Multilayer printed boards for wiring have already been shown to be possible when used with insert molding (Peltola, 2006; Alajoki et al., 2009), which serves to highlight the possibility inherent for further multilayer electronic structure integration.

Other applications include in-mold lighting and identification labels both of which are also added retrospectively to plastic molded parts. Middleton and Goodship (2012) reported development of new method to incorporate electroluminescent (EL) function during manufacture (see device in Figure 4.23). This involves applying a multilayer EL structure into an injection mold cavity using a spray deposition process, before back injection to produce both a molded plastic part and a functional EL surface. This application utilized commercially available conductive polymer poly(3,4-ethylenedioxythiophene) doped with poly(styrenesulfonate) (PEDOT:PSS).

Radio-frequency identification (RFID) technology presents potential to both trace appliances and their usage as well as collect data relating to plastic products. Such a system utilizes radio waves to transmit data from a tag placed on or within the product. These come as either active (require a power supply) or passive devices. There are design advantages and disadvantages to both types that are beyond the scope of the discussion here.

Direct integration of printed conductive structures into plastic parts was developed specifically for identification applications. Like the EL structures this application also utilized the conductive polymer PEDOT: PSS, and the memory structure was printed by flexographic printing. Like similar IML techniques, the printed films were cut into labels and fixed into the mold, followed by back surface injection molding.

The functional structures in this case consisted of one basic electrode and several memory electrodes and were detected using capacitive near-field coupling. This works by applying a voltage at the input electrode and then detecting a signal at the output electrode. It is reported functional samples were produced, but it is also reported that high temperatures, pressures, and injection speeds could cause degradation and distortion of the electrical structures. In this case, melt temperature up to 280°C and mold temperatures of 80°C were used. A patent was granted on in-mold RFID labels in 2006, and is now commercially available (Inotec, 2012).

Other potential multifunctional structures are touch switches and control panels. Their production using injection molding can be found from patents that date back to before 2002; however, at the time of writing there appears little information on products in the marketplace that actually use this technology.

The sensors can be incorporated into flat or curved surfaces to replace mechanical switches sliders or wheels. The result is a single mold solution that is durable, chemical, and water resistant. As capacitive touch sensors can be integrated within plastic products again by using the IML method, this technology is seemingly desirable for equipment and controls, particularly in the electronics, mobile communication, and automotive sectors. Further IML uses include copper circuits that are preprinted onto the film substrates that can be subsequently inserted into the mold followed by injection molding to create a robust functional interface on the component.

Future developments include more complex solid state systems, for example, the integration of photovoltaic films and functional batteries into injection molding. While still at the research stage, market development makes these kinds of electronic products likely to appear within the next 5 years for use in mold.

4.4.1.7 Biinjection Methods

When multiple injection units are aligned into one mold the process of biinjection becomes possible, but it is necessary for the tool to essentially change in some way between shots, for example to rotate the tool or for sliders to change the cavity. There are two options for this process that can either be controlled by the injection molding machine by the use of an indexing plate or the tool itself can be built with rotatory capability.

It is necessary for the rotation to be very precisely controlled, while also remaining economically viable with realistic speed of stroke and rotation. The effect of this movement must also be cushioned enough to soften the impact of this rotation of the tool and machine. Machine-based systems use an indexing plate on a moving rotary platen to achieve this accuracy. It is still necessary to incorporate normal injection components such as the integration of ejector pins. While for regular use this provides a longer term economic advantage for purposes where there is not regular use it may be preferable to buy a more expensive tool that has the same built-in features.

Materials are both processed and cycled simultaneously and can be used to build up multishot components as shown previously in Figure 4.3 with a shot being produced in each cavity at each rotation point. The entire cycle is, therefore, controlled by the component (and cavity) requiring the longest cycle time in this process.

There are a number of different ways this can be achieved as shown in Figures 4.24 to 4.26. In one option the tool moves to expose different cavities. This can be achieved with sliding cores as shown in Figure 4.24.

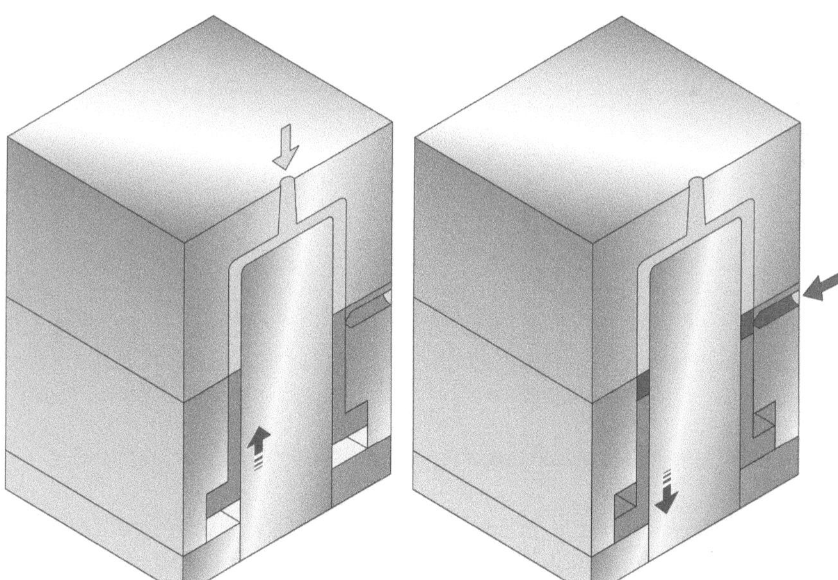

Figure 4.24 Cavity is closed and opened using mold design with core puller activation.

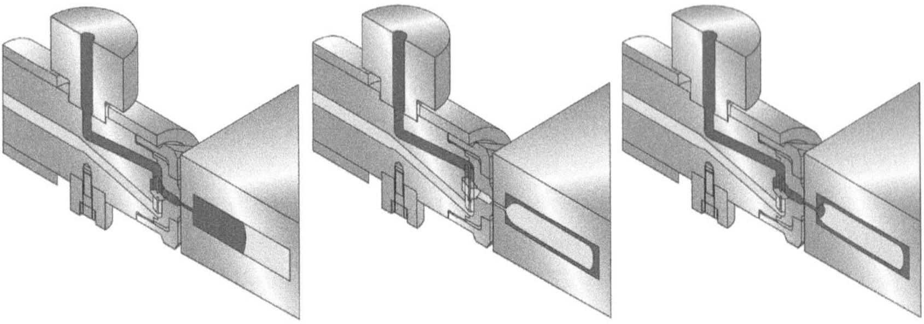

Figure 4.25 Mold filling using two component feeds.

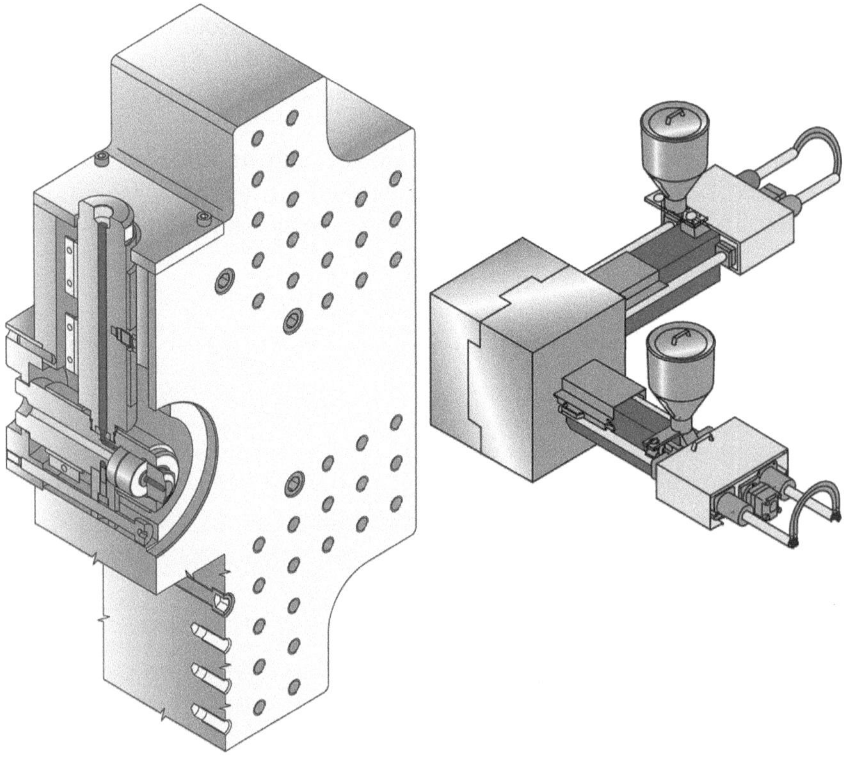

Figure 4.26 Manifold to switch feeds (left), machine configuration (right).

One part of the molding is injected and then the tool is activated to expose a second cavity into which the second component is injected.

A further method is to use a technique whereby the material is injected into the same cavity at the same access point, through the use of a manifold system to switch the two feeds on and off as shown in Figure 4.25.

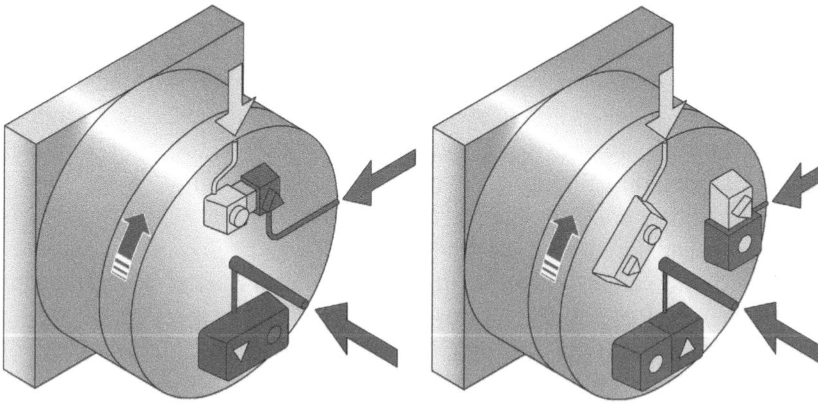

Figure 4.27 Examples of three-step rotation.

A manifold is shown in Figure 4.26 and an example machine configuration is shown on the right in the same figure.

The use of a rotating core offers even more possibility as shown in Figure 4.27. Here, three different feedstocks can be molded to make up as the tool is rotated by 120°.

A further example of a rotating tool is shown in Figure 4.28. This kind of tool is known as a cube tool and turns 90° at each cycle and can be used to produce the multicolor packaging components such as lids.

4.4.1.8 Multishot Options

As touched on briefly earlier in this chapter, robotic solutions have provided manufacturers with a variety of options for automated and accurate control of part placement, registration, and removal of in-production components.

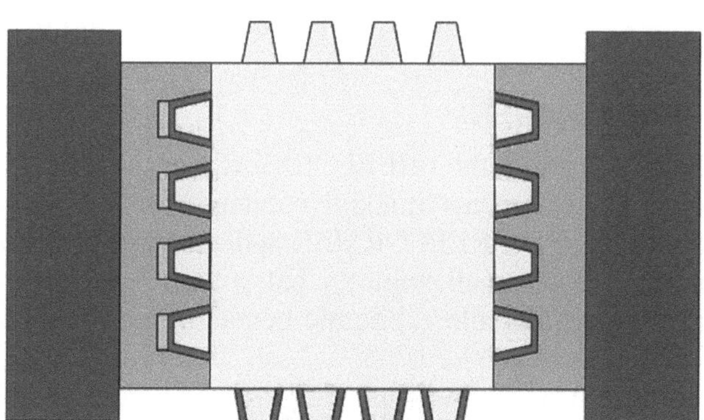

Figure 4.28 Cube tool.

There are a number of potential options and configurations for multi-shot products that will now be covered.

With a single machine, rotation can be built to enable multiple cavities to move and be indexed either by the tool or the machine using an indexing plate. In this way, individual cavities can be allowed to cool, be overmolded, or ejected as required. A variation on this technique is to allow a robot to move the components from one tool to another when the tool is open, rather than rotation of the tool itself.

If there is no sound reason to invest in a rotating plate such as an indexing platen or it is simply not possible to use one, then robots allow the same function of moving and placing components into different areas of the same tool or into the tool of a different machine.

4.4.1.9 Robotic Direct Repositioning in Mold

As there is a choice of configuration here for both tooling and component design, repositioning in the mold can be considered either insert molding or over molding depending on the particular process being used. A robot can be used to reposition a component in the same mold (eliminating the need to index) or transfer that component to another mold in a second machine (also eliminating any need to rotate the tool). Robots are already heavily used in overmolding to place inserts with great accuracy again and again.

4.4.1.10 Hybrid Processing Systems

As well as IMD advances, hybrid processing systems that incorporate elements of other processes are also appearing on the market place, for example, 3D printing has become used not just for prototyping but also for producing short volume runs for as prototype tooling. Further, as the boundaries between processes blur, ARBURG have produced a hybrid 3D/injection molding system in the recently launched FREEFORMER.

4.4.1.11 FREEFORMER

At the time of writing, the FREEFORMER was not yet fully launched but is set to combine elements of additive manufacture (see Chapter 5) and injection molding. The FREEFORMER will potentially provide highly efficient production of small volumes but without the need to invest in tooling offering considerable economic benefit given the cost of tooling solutions.

The trend pushing this development comes from a combination of the freedom 3D printing provides in design but coupling this with wider material choices for production seen in injection molding. The machine is

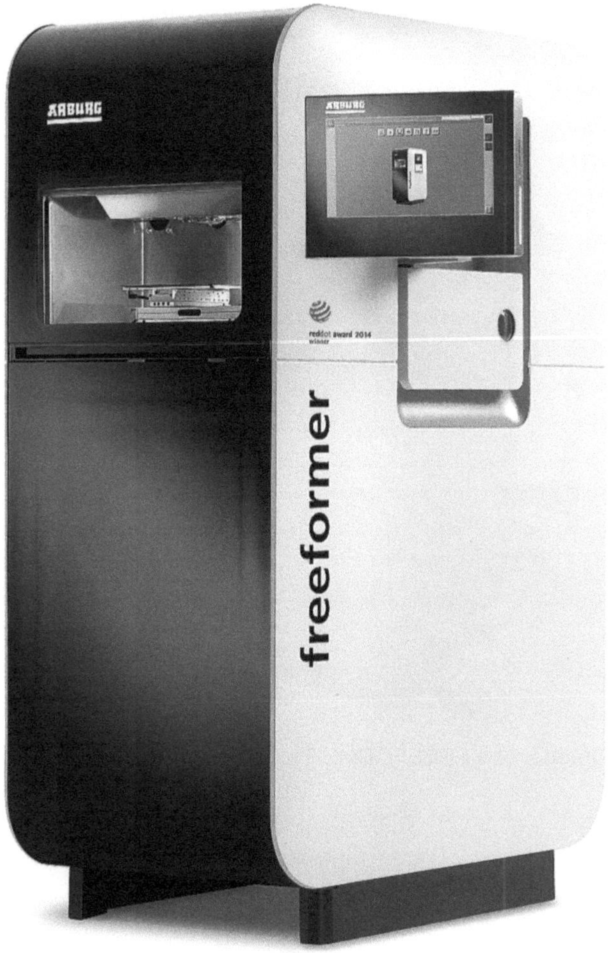

Figure 4.29 FREEFORMER machine by ARBURG. PHOTO: ARBURG.

shown in Figure 4.29, the interior is shown in the schematic in Figure 4.30, and a part made by the machine in Figure 4.31.

With the advantage of utilizing pelleted feedstocks and the fact high temperature engineering polymers, the materials generally considered "more difficult" for injection molding can be formed, the FREEFORMER offers material choices not currently available to traditional additive manufacturing methods. The realized properties are about 80% of those that can be achieved by injection molding. This is the result of the different forming method and is comparable with similar additive manufacturing methods. An example part is shown in Figure 4.31, which utilizes hard

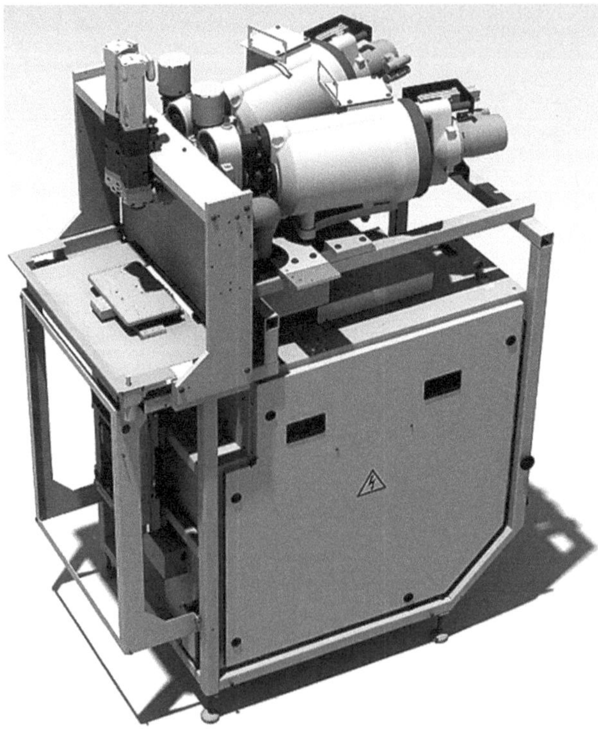

Figure 4.30 Interior of a FREEFORMER machine.

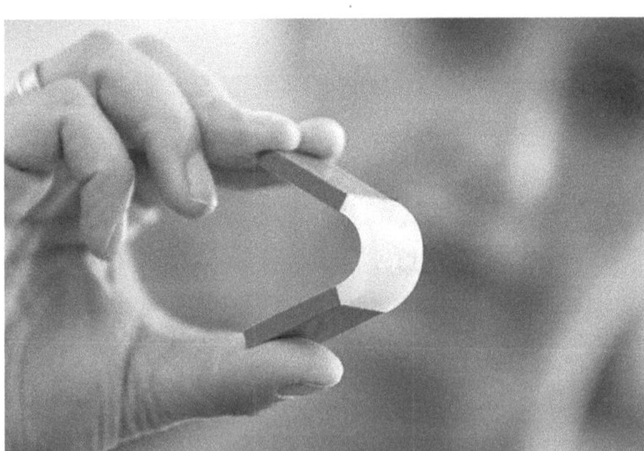

Figure 4.31 Part made from FREEFORMER using soft and hard material combination.

and soft materials to produce a part that would be extremely difficult to produce by injection molding.

4.5 Application Considerations

4.5.1 Automation and the Internet of Things

Industry 4.0 was a high-tech strategy adopted by the German government in about 2012. The aim of this strategy was to develop the smart factories of the future, creating linked intelligent factories by mass computerization. The major characteristics of these idealized factories would be adaptability, customization, and integration that linked back through the supply chain customers and forward to their business partners.

This development is often termed the fourth industrial revolution and is expected to be a reality in about 10–20 years powered by a growing interest in linking cyber systems powered by concepts such as the "Internet of Things" which are the enablers that will continue to bring such systems into reality.

Similar themed initiatives under different names such as Factory of the Future (UK) and Smart Manufacturing (United States) are found elsewhere around the world in government and company strategies, which all envision the powers of automation and internet enablers to create a multi-stakeholder vision of future factories and systems. For injection molding this has been seen as a growing trend to link all automation systems together through a single machine interface in the injection molding machine that can be externally linked to program or transmit production information. Figure 4.32 shows an automated conveyor for production components to be fed away from the machine; this could equally show other systems already introduced such as robot pick and place machines or to include automatic quality checks to create complete manufacturing cells incorporating preprints for IML, or inline on demand printing, component molding, robotic switching between machines to incorporate added multifunctionality, robotic part removal and nondestructive inspection and quality control functions along with complete packaging solutions ready for dispatch. All controlled and fully integrated by the injection molding machines at the heart of the process.

These trends can already be seen across many sectors; however, for the purposes of the brevity of this book only two application sectors are considered here: packaging and electronics.

Figure 4.32 Machine side conveying system.

4.5.2 Packaging

Plastics are widely utilized by the packaging industry due to the inherent advantage over other materials most notably their flexibility when compared to glass as well as being of lighter weight than metals; they are also easily formed into a variety of shapes. The packaging sector is very innovative, and is just one field where the advantages in cycling speed that electric-driven injection molding machines can be utilized in high volume and competitive cap and closure market. One packaging example freely available on the internet shows 72 screw top caps produced in a multicavity mold in just 3 seconds!

Electric machines can also allow the improvement of related factors such as energy efficiency and carbon emissions in some cases (see section on energy efficiency).

Of most relevance to injection molded applications is the production of closures such as caps and lids for the food industry. A variety of tubs and bucket-like shapes are also produced. A further use, which is generally dealt with as an integrated and distinct process, is the production of the preforms that go onto produce injection blow molded bottles.

Depending on the specific application, design considerations for packaging can include a number of important material properties that may include density, impact strength, moisture and gas permeability, chemical resistance, tear strength, heat sealability, heat resistance, and whether it can be sterilized for food and/or medical applications. Optical properties for customer aesthetics may also be of utmost importance in some cases. As part design becomes more complex with the use of undercuts for example, part ejection becomes much more complex and tool design must be more complex, for example with the case of parts that cannot be popped loose directly it may be necessary to mechanically unscrew them from the mold core or have a collapsing core that retracts.

One advantage the packaging industry has embraced is in mold decoration and labeling with the entire process through to packing integrated into production cells through the increasing use of automation. Intelligent labeling solutions incorporating consumer guidance on freshness, for example, will be more widespread in the future. Of further interest to the packaging industry in the future is the potential to link products to the wider development of internet of things, by for example ordering fresh milk when the packaging registers it is empty. Advances in IML, such as the use of freshness sensors enabled by IME, also give designers plenty of opportunities to innovate in the future.

4.5.3 Electronics

There are two important technological drivers in the electronic industry. One has been the effect of the European Directive for Waste electronic and electrical equipment (WEEE Directive), the other is the development of new technologies, specifically plastics electronics. Where the plastic industry mainly provided cases and containers for electrical goods, they are now becoming part and parcel of electronic equipment. Further, longer term concerns over the stability of supply of some rare metals have also led to a search for alternative materials for the future. Plastics are predominantly used for their insulating properties; however, with certain additives they can be made more conductive as discussed in Chapter 2. Further developments in conductive polymers and interest in conductivity enhancing materials such as graphene and carbon nanotubes will also be used more widely in the future. Due to their transparency, flexibility, and ease of manufacturing plastics are increasingly gaining markets as the substrate materials of choice for the electronics industry. Several examples were given previously including RFID and electronic injection molding.

4.6 Impacts

There are a number of complex and sometimes opposing factors that the injection molding industry is subject to, such as economic, political, legislative, and financial constraints that can have local and global impacts and related knock on effects. That is without the further consideration of the impacts of consumer power and consumer trends and ongoing societal interests more widely. While it is not also possible to separate and isolate factors, some consideration to individual effects is given here.

4.6.1 Legislation

Legislation has already been shown to be an ongoing and developing important driver toward sustainability, impacting the plastics industry through legislation imposed in turn on the packaging, automotive, and electronic industries around waste, producer responsibility, and recycling targets.

4.6.2 Cost/Finance

As will also be discussed in the section on energy efficiency, the way a company approaches the business and its longer term investment strategy or shorter term targets and goals can have a major impact on the potential to invest and integrate new technology. For example, a company in the USA may be far more interested in investing in and driving innovation to protect market share, than a company in Nigeria looking to keep investment costs down rather than update the capability.

4.6.3 Environment and Disposal

Driven largely by legislation emanating and developing out of Europe since the 1970s, there has been a slow but increasing trend to consider how a component is designed and assembled and also the knock-on effect on how it is disassembled and recycled. As environmental impacts have become more widely known, accepted, and understood, a more complex cradle to grave approach to design is now considered more appropriate for new product design.

This approach has come out of ongoing lessons learnt by successive introduction of legislation such as the Waste Framework Directive (Directive 2008/98/EC on waste), The Directive on End-of Life Vehicle (Directive 2000/53/EC) and as technology develops, a very fast growing and important newer waste stream, that of waste electronic and electrical equipment (Directive 2012/19/EU), known as the "WEEE directive."

Going further still, a Green paper (meaning suggestions on how a law may work but not yet law) was issued in 2013 on a "European Strategy of Plastic Waste in the Environment." This means new more wide ranging legislation is likely in future years.

For readers interested in further information the European Commission website contains a variety of data on legislation, collection rates, and recovery costs across Europe. The environmental section of the website can be found at http://ec.europa.eu/environment/index_en.htm (accessed March 2014).

4.6.3.1 Recycling

While plastic recycling has become common place (Figure 4.33), one problem with increasing multifunctionality is as parts become more complex, their recycling and separation also become a more complex issue. There is therefore always going to be a tradeoff between the complexity of the product (and therefore process integration and associated energy and cost savings) and the need to provide a cradle to grave sustainable and environmentally sound solution. However, products designed with this in mind from the off-set can still be produced that achieve both. It is also necessary

Figure 4.33 Recycling (pictures freely available from RECOUP).

to consider that it is environmental legislation that drives the related search for practical and sustainable disposal solutions.

4.6.3.2 Life Cycle Analysis (LCA)

With the legislative mechanisms likely to increase in future and environmental responsibility unlikely to come off social, political, or world agendas, life cycle analysis is increasingly important to manufacturers who are no longer just responsible for a product but also increasingly for the product's entire life cycle. LCA is a methodology to aid assessment of anything from a service, product, or process and involves compilation of inventories of all associated energy uses, material or process inputs, and subsequent releases into the environment. By evaluating all these individual inputs, it is possible to gather information of the total environmental burden of the system. From here, it is possible to undertake comparisons of different materials, designs, and process routes to enable an entire cradle to grave approach to the life cycle of any product. For example, the impacts of individual process routes versus the kinds of integrated processes described here can be considered in this case not just for economy but for the likely impacts on the planet and society.

4.6.4 Society

4.6.4.1 Energy Efficiency

Energy savings, examining our carbon footprint, and examining our lifestyles to obtain more low carbon lifestyles have affected all aspects of society from energy-efficient light bulbs to growing recycling schemes. The change to a recycling society rather than a throw away one has fundamentally affected the way we as a society behave. It is, therefore, not surprising these same drivers are present in a more commercial setting.

One of the advantages of incorporating multifunctionality in the injection molding process is the potential for increased cost savings and energy-efficient production. These competing factors are shown in Figure 4.34. The high-energy usage of polymer processing factories and equipment mean there are considerable benefits and opportunities in adopting energy-efficient practices and there are a number of recommendations for energy-efficient practices in polymer processing readily available online (i.e. http://www.tangram.co.uk/TI-Energy%20Worksheets%20%28Plastics%29%20-%20Tangram.PDF accessed March 2014).

However, these benefits are often not considered in light of the other considerable advantages they provide. Therefore, these barriers are both identified and considered here to stress the effectiveness of an energy

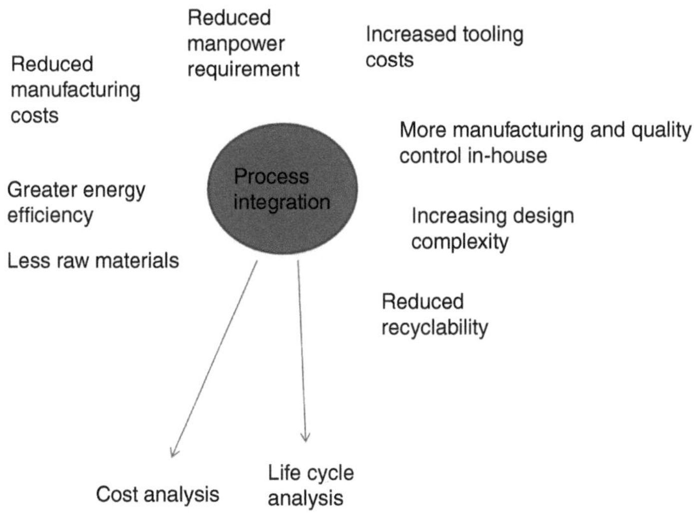

Figure 4.34 Competing factors in multifunction.

reducing and profit increasing energy management plan. However, the adoption of energy-efficient practices is not driven by a simple desire to be more energy efficient; often there are other more complex economic, political, and organizational factors to consider.

Therefore, defining the factors is equally important whether it be economic such as an overall limitation to capital investment as a whole or decision is constrained by limited finances. Studies in Greek and Swedish firms (Fleiter et al., 2012) confirm that the availability to capital access is the largest barrier for investment in this area, with the investment priories of each company (Thollander and Palm, 2013) and their aversion to borrowing capital, a fact constraining developing countries as well (Alcorta et al., 2014).

The concepts of "offensive" and "defensive" spending are worth considering here, where certain areas of investment, such as marketing (to increase sales), are more attractive than cost cutting activities (such as energy efficiency). It was found, for instance, that around 80% of German firms prefer to invest in areas that relate to their core business targets rather than energy efficiency. Accountancy and payback of investments can also lead businesses away from energy efficiency targets (Fawkes, 2013).

4.6.4.1.1 Lack of Adequate Information

Despite the seemingly large amount of available information on improving energy efficiency performance, the lack of information surrounding management (including financial) potential and technology for energy efficiency savings is still observed even in more developed countries and industries (Fawkes, 2013; Fleiter et al., 2012).

The mindset of each and any company will affect how it behaves; factors such as organizational barriers such as culture, structure, lack of the required skills, and internal capacity. Behavioral barriers are also present around credibility and trust of available information sources, lack of environmental value, and a simple technological inertia due to a resistance to develop and change. This can be considered a reluctance to buy new machines and periphery devices until forced to by breakdowns, etc., despite being aware of the benefits of energy efficiency. It is these kinds of difficulties that inhibit development within the industry.

Satisfactory identifying and dealing with the uncertainties and the risk associated with any kind of investment constitutes a significant barrier to development in injection molding industry and generally any introduction of more energy-efficient systems within industry is necessarily associated with risk.

4.7 Barriers to Multifunctionality

4.7.1 Machine Design (Screw Processing)

The limit on multimaterial processing in injection molding is that this is a screw-based processing system and therefore materials need to pass from the hopper through the screw and be injected into the mold in some manner. This can only be bypassed by insertion directly into the mold cavity either before or after an injection stage has taken place. Clever tool designs and multisteps can be used to overcome this to some extent but at the heart of any injection molding process is the need to meter and inject through a screw process. Therefore, there are the following limitations on the process: materials need to operate at high shear, with built-in process orientation, be heat stable, rheological suitable, as well as keeping process temperature, pressure, and cooling gradients to a minimum wherever possible.

4.7.2 Materials

While there are a large number of materials available for injection molding, the use of some materials combinations is limited by any number of issues related to adhesion, wettability, incompatibility, rheological instability, cure times, heat stability, and heat transfer. Therefore, it is paramount that consideration be given as to the suitability of the use of certain multimaterials systems in the design stages.

4.7.3 Tooling Costs

As tooling complexity increases so the capital cost of tooling increases substantially. Therefore, it is of utmost importance that cost comparators are used such as provided in Table 4.1 to ensure economic viability of process integration methods.

4.7.4 Heat Transfer and Cooling Complexity

As previously stated in Section 4.3.5, anything in the mold that limits the ability of the tool to heat and cool the component will have an effect on the most important component of the injection molding cycle, that is the cooling time. If using dissimilar materials there is also the possibility that cooling times will vary across different parts of the molding building in further nonisothermal cooling.

As material systems become more complex, and different parts of the final components have different thermal and conductive properties, modeling the cooling effectively becomes more difficult, cooling times may become longer due to the use of increased insulation in the design, and production savings from integration become lost as cycle times must be increased. There is, therefore, a tradeoff here between the complexity versus economics relating to increases in the cycle time and material performance in terms of increasing molded in stresses due to different material characteristics.

4.7.4.1 Recycling/Legislation

As parts become more complex, their recycling and separation also becomes a more complex issue. There is, therefore, a tradeoff between the complexity of the product and therefore process integration and associated energy and cost savings and the need to provide a cradle to grave sustainable and environmentally solution. However, products designed with this in mind from the offset can be produced that achieve both.

4.7.5 Advantages Versus Disadvantages

Table 4.13 compares injection molding against two other competing multifunctional processing techniques. Thermoforming is included here as it uses sheet feedstock and can use multilayered films to produce multilayered films that could also therefore in the future be used to make more multifunctional products. It can be seen that for low production runs there are advantages in using other processes within their own process-specific constraints; however, for high volume and fast production of thermoplastic

Table 4.13 Comparison of Processes

Process	Composite Techniques (C) (Compression Molding)	Injection Molding (IM)	3D Printing (3D)	Thermoforming (T)
Materials	More limited than IM	Wide range of materials	Limited material range	Sheet feedstock needed but could be made of multimaterials and therefore functional layers
Production speeds	Slow in relation to IM but complete parts possible. IMD possible	Very rapid, complete parts possible. IMD used	Slow build speeds, part may need further work to eliminate support materials	Slow but multiple components may be produced at the same time – needs further trimming
Tooling	Cheaper than IM	Expensive tooling	No tooling	Very cheap tooling
Accuracy	Better than thermoforming	Excellent accuracy	Less then IM	Less accurate, limit design
Production runs	Good for short runs	Not good for short runs	Designed for single run	Good for short runs
Surface effects	Some possible	Many surface effects possible	Surface effects limited by machine resolution	Surface effects limited
Designs	Less than IM	Complex design but less than 3D printing	Complex designs not achievable by other processes but accuracy less than IM.	Undercuts restricted

Process integration	Possible: such as injection-compression	In mold labeling possible, advanced processes developed	Limited	Limited
AUTOMATION	YES	YES	NO	YES
Mixing	No mixing	Good material distribution	No mixing stage	No process mixing possible
Orientation of fillers	Can be built in or built out as required in design	Natural orientates materials in flow direction and therefore can be problematic in some products	No orientation possible	Depends on the feedstock used
Costs per part: indicative only – will depend on product and complexity				
Low production < 100 parts	High	Very high	Low	Medium
Medium production < 1000 parts	Medium	High	Medium	Medium
High production > 5000 parts	Medium	Low	High	Medium

materials requiring accuracy, complexity, and precision, there is no competition to the injection molding process. Only within specific design constraints such as:

- compression molding with its ability to control the fiber orientation for specific high-strength composite materials and therefore produce materials beyond the possibility of injection molding designs
- and 3D printing with its ability to produce complex interior designs not currently possible with injection molding

It is only under these kinds of circumstances that viable alternatives to the use of injection molding present themselves. As advanced injection molding processes continue to be developed and new hybrid machines introduced, such as the FREEFORMER machine by ARBURG, it is unlikely that these advantages will be eroded in the foreseeable future.

4.8 In Conclusion

The advances in injection molding processes, design, and products have led to the introduction of more and more complex and multifunctional products with increased design freedom, mass customization processes, and process integration. While the environment may be the ultimate beneficiary on the drive toward process efficiency, at the heart of all advances is a fundamental need to attain economic profitability for the manufacturers. While injection molding continues to adapt and maintain this potential for improvements in manufacturing efficiency and production cost savings, it will retain its position against other process routes in the future. It is likely further hybrid production routes that combine elements injection molding lack such as orientation control (such as in compression molding) and interior design control (such as 3D printing) will follow in the future.

This review has provided an insight into the potential to integrate new functionality to parts within the mold using a variety of processes and process routes. IMD and IML techniques have evolved to provide more than a decorative surface and new applications have been investigated including the successful production of injection molded parts with integrated lighting capabilities, RFID tags, and capacitive films. The technology enables designers and manufacturers to meet market demand by developing products with increased functionality while reducing the number of processing steps. Utilizing this technology could ultimately open up the opportunity for other multilayer components such as photovoltaic cells and batteries.

The influence of initiatives such as Factory 4.0 means the complete link-up and remote automation of manufacturing systems. This included fully customer-customized orders, through the procurement of raw material and linking through the entire supply chain. With inclusion of automated production, quality control, and logistics already, this will also expand to include not only tracking a product through a product life but also until ultimate disposal, as multifunctional products not only become more and more routine but also ultimately allow the materials within them to be more traceable. Further reading can be found at end of book.

References

Alajoki, T., et al., 2009. In-mold integration of electronics into mechanics and reliability of overmolded electronic and optoelectronic components. In: European Microelectronics and Packaging Conference (EMPC), 15–18 June, Rimini, Italy, pp. 1–6. IEEE.

Alcorta, L., Baziliain, M., De Simone, G., Pedersen, A., 2014. Return on investment from industrial energy efficiency: evidence from developing countries. Energy Efficiency 7 (1), 43–53.

Fawkes, S., 2013. Energy Efficiency: The Definitive Guide to the Cheapest, Cleanest, Fastest Source of Energy. Gower Publishing Limited, Surrey, England.

FBT, 2015. <http://www.fpt.de/de/produkte-leistungen/digitaldruck_GB.pdf> (accessed 01.13.15.).

Fleiter, T., Gruber, E., Eichhammer, W., Worrell, E., 2012. The German energy audit program for firms: a cost-effective way to improve energy efficiency? Energy Efficiency 5 (4), 447–469.

Inotec, 2012. RFID in-mold labels. <http://www.inotecbsl.com/products/rfid-labels/rfid-in-mould-labels/> (accessed 01.03.15.).

Lentz, A.M., 2010. In-mold decorating with pre-printed film inserts. Mod. Plast. Plast. Decorating Mag., 20–23, (November).

Love, J.C., Goodship, V., 2002. In-Mould Decoration of Plastics. Rapra Technology Ltd, Shawbury, UK.

Makenji, K., et al., 2006. A new in-mould decoration process using real wood veneer. Progr. Rubber Plast. Recycling Technol. 22 (4), 225–242.

Middleton, B., Goodship, V., 2012. Injection molding electroluminescent components. Polym. Eng. Sci. 2012, 1–9.

Owen, P., 2011. Energy saving trust, the rise of the machines: a review of energy using products in the home from the 1970s to today. <http://paulaowenconsulting.co.uk/RiseOfTheMachines.pdf> (accessed 01.03.15.).

Peltola, T., 2006. Integration of Multilayer PWB into Plastic Covers by Injection Moulding. 1st Electronics System Integration Technology Conference, 5–7 Sept. IEE, Dresden, Germany.

Thollander, P., Palm, J., 2013. Improving Energy Efficiency in Industrial Energy Systems: An Interdisciplinary Perspective on Barriers, Energy Audits, Energy Management, Policies, and Programs. Springer, New York.

Further Reading

Fleiter, T., Worrell, E., Eichhammer, W., 2011. Barriers to energy efficiency in industrial bottom-up energy demand models – a review. Renewable Sustainable Energy Rev. 54 (4), 3099–3111.

5 Multifunctionality in Additive Manufacturing

Benjamin M. Wood

5.1 Introduction to Additive Manufacturing

Few areas of advanced manufacturing and engineering have received the same level of media interest as additive manufacturing (AM). The global market for products and services in this sector is growing rapidly, worth over $3 billion in 2013 (Wohlers, 2014) and forecast to exceed $100 billion annually by 2020 (Wohlers, 2012). Despite the recent growth in interest, AM is not a new technology and was pioneered in the 1980s, with selective laser sintering (SLS), stereolithography (SLA), and fused deposition modeling (FDM) all commercialized by 1990. Today, the majority of patents surrounding these technologies are owned by two large companies from the USA; Stratasys Inc and 3D Systems. Through mergers and acquisitions they have grown to represent the majority of the market in Additive Manufacturing systems, and at the time of writing collectively employ more than 4000 people worldwide.

Since 1990 other companies have carved their own niche by developing unique processes that are not protected by intellectual property rights, and are often used for more specialist applications not best suited to the three main technologies listed above. There has been significant innovation in this space and there are a huge variety of processes, materials, and properties available.

Due to the maturity of AM technology, however, many of the patents protecting these pioneering technologies have now expired, with the result that many smaller companies have been created over the last 5 years with their own interpretation of the traditional AM processes. The printer credited with being the first of the current crop of low-cost AM machines is the "RepRap" (short for **Rep**licating **Rap**id-Prototyper), developed as part of a research project to create a self-replicating machine (Jones et al., 2011). The RepRap is built from parts that can either be made on another RepRap or are cheaply and readily available such as fasteners, threaded bar, and simple electric motors. It makes use of the FDM process, although due to trademarks this is described as fused-filament fabrication (FFF). Due to its open hardware license, this concept has been copied, modified, and developed since its launch in 2004 into many

Design and Manufacture of Plastic Components for Multifunctionality
http://dx.doi.org/10.1016/B978-0-323-34061-8.00005-3

hundreds of low-cost AM machines, most of which are purchased by hobbyists and small businesses.

What all of these systems, both high- and low-cost, have in common is that they construct a component by repeatedly melting, solidifying, or curing thin layers of material. By "slicing" a 3D component into a succession of layers the manufacturing process can be simplified into 2D movements, meaning that AM can produce incredibly complex parts or even assemblies with ease.

We are living in an exciting time for this technology, with significant opportunities for well-established AM techniques to be further developed, adapted, and improved by researchers, hobbyists, and entrepreneurs. This provides plenty of opportunity for innovation, and with AM this inevitably means multifunctional materials and parts will become more widespread.

In this chapter, some basic theory of the processes and materials behind AM will be introduced, followed by a look at how multifunctionality can be achieved with existing commercially available technologies, and finally what tomorrow's AM machines will be able to achieve.

5.2 Background and Terminology

The subject of this chapter goes by many different names and aliases depending on who you speak to. The more astute reader may already have noticed the large number of acronyms used in describing these technologies. For reference, a list of these acronyms is included at the end of the chapter.

The mainstream media love the term 3D printing – it is easy to understand and reasonably descriptive of the processes used, but can potentially mislead us into thinking that making a component is as simple as clicking the "print" button on our office photocopier.

Those who used the technology in the past may use the term rapid prototyping (RP). In the late 1980s and early 1990s, this accurately reflected the only real use for parts made using this process; it was a faster and sometimes cheaper way of making initial models to check form and fit. Although this application of the technology is still widely used, advances in software, process, and materials mean that users are no longer limited to making prototypes, and functional parts can now be manufactured. For this reason, RP is no longer widely used to describe this group of technologies.

Most purists would probably prefer additive manufacturing (AM). AM refers to a process by which digital 3D design data are used to build up a component in layers by depositing material; this definition was presented by an ASTM International Committee in 2012 to prevent confusion and ambiguity between the various terms being used both within industry and in the media (ASTM, 2012). This definition is now the one adopted by industry and reflects the huge range of different manufacturing processes and variations of machine that have been developed in recent years, and is the most all-encompassing.

Whichever term you prefer to use, what is certain is that there has been a huge growth in the use and awareness of AM in recent years, with the result that there are now more machines and processes available to potential users than ever before. From printers costing just a few hundred dollars to multimillion dollar machines capable of manufacturing parts in precious metals, there is a system available to suit most requirements.

The growth in awareness of AM has inevitably been driven by the media, both traditional and social, and therefore there is a need for a certain amount of expectation management before getting into a deeper discussion about this interesting topic. Some articles would have you believe that AM is the "magic bullet" for manufacturing; that in 10 years everyone will have a 3D printer in their home and that we will never need to buy consumer products again because we can make them in seconds in our living rooms. The reality is somewhat different.

As with more traditional computer numerical control (CNC) machining, building a part using AM requires a number of steps to be followed that vary from process to process but follow a similar theme.

5.2.1 The Additive Manufacturing Process Flow

Any AM component begins life as a digital design, usually generated using computer aided design (CAD) software, but also potentially generated using scanning technologies such as laser scanning or photogrammetry. The vast majority of AM systems require files to be uploaded in .STL format – this stands for **ST**ereo**L**ithography and was introduced by 3D Systems in the late 1980s specifically for their stereolithographic AM process (Figure 5.1). This file is a polygonal surface mesh of a solid object and so does not contain any information relating to color or texture.

The STL file is imported into another piece of software that processes the part ready for building on the desired AM machine. It also "slices" the part into 2D layers that can be manufactured sequentially by the AM machine. This stage can involve definition of parameters such as part orientation, the material used for manufacture, and the layer thickness (which in turn defines the "smoothness" of the finished part). The properties of parts made using some AM processes are orthotropic; because the part

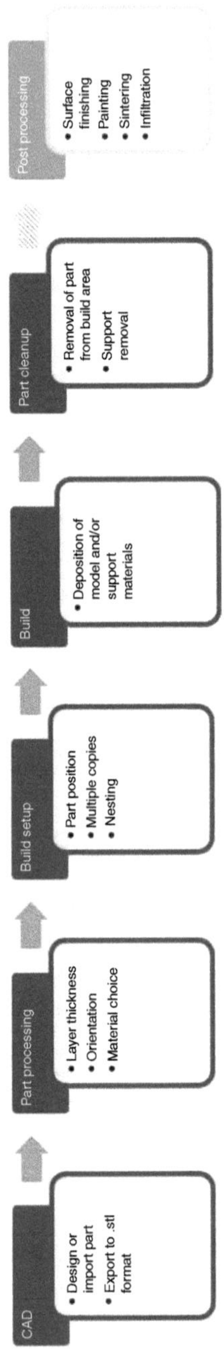

Figure 5.1 Additive manufacturing process flow.

is built layer by layer, the structure is inevitably weaker when loaded perpendicular to the layers. This is particularly true for filament extrusion processes; therefore, consideration should be given to the direction and type of loads that are likely to be applied to a part during use, and the orientation should be chosen accordingly.

The sliced file is then exported to another software package linked to the AM machine that enables the user to position multiple parts onto the build area, including copies of the same part if desired. By building more than one part at the same time, the overall time-per-part is reduced, therefore reducing overheads and costs.

Once the build has completed, the part or parts have to be removed from the machine that may involve removal of excess resin or powder, and then any support structures must be removed. For some systems this will require manual removal using tools; for others the parts may be left in a solution of material that dissolves the supports and leaves the part untouched.

Finally, there are a variety of processes that can be used to improve the surface finish and appearance or even the mechanical properties of the parts postprocess. These are usually optional but depending on the system used postprocessing may be required to achieve the best mechanical properties. An example of this is metal powder systems that use an energy beam; without heat treatment of the parts postprocess, the internal stresses generated by the build process can be extreme and may result in part failure.

AM therefore requires a sequential series of processes of some complexity, including access to CAD or digital design software, and therefore the learning curve for anyone new to the process can be daunting. The major benefit is that this process flow enables highly complex parts to be manufactured with no greater difficulty than very simple items.

5.3 Additive Manufacturing Processes and Materials

The nature of intellectual property in AM means that numerous ingenious processes have been invented to carry out the layer-wise deposition of material without violating patents. One of the results of this is that material and process are almost inseparable – the two are typically developed in tandem to produce a unique process with its own range of material options and functionalities. In this section, some of the more common processes and their accompanying materials are introduced, before some of the more innovative research concepts of recent years are described.

5.3.1 Extrusion Processes (FDM/FFF)

Fused deposition modeling is now probably the most widely used and available AM technology. It involves the extrusion of a polymer filament through one or more heated nozzles that are controlled by three axes to build up the part layer by layer. In concept this process is the closest to traditional CNC machining or subtractive manufacturing; in most systems the same programming language (G-Code) is used to control the movement of the axes. The outline of each layer is first deposited; then, the enclosed area is filled using a repetitive pattern that can be varied to create different internal structures and therefore vary the mass or strength of the finished part. This process is repeated until the model is completed. Parts are built on a rigid plate that is sometimes heated to improve adhesion of the first layers. In addition "raft" structures are sometimes used in the first one or two layers to prevent narrow structures with high aspect ratio from coming loose from the build plate and therefore causing a failed build.

All except the very low costs systems use more than one extrusion nozzle in order that a "support" material can also be deposited, allowing overhanging or hollow parts to be produced. This can either be made from the same material as the rest of the part, in which case it is normally given an open lattice structure to make it easier to remove from the finished part, or from a specialized support material that can be removed without the need for tools. Dedicated support that can be chemically dissolved requires less labor and enables removal from hollow parts with limited access for mechanical removal methods.

Almost exclusively, these systems build using thermoplastic filament that can be made from a huge range of plastics including acrylonitrile butadiene styrene (ABS), nylon, polyetherimide (PEI), polyphenylsulfone (PPSF), and polylactic acid (PLA). These range from biobased polymers to engineering plastics that will withstand attack from most solvents and chemicals, to those that will function at the high temperatures found in a vehicle engine bay. Some approximate properties of commonly used thermoplastic filaments are included in Table 5.1. These values should be used as a guide only; they are averages taken from datasheets of different suppliers to allow comparison. Each filament will have different properties depending on the exact grade of polymer and the processing parameters used in its manufacture.

5.3.1.1 Support Removal

Caustic baths or wash tanks are often used with parts made from FDM to chemically remove the support material. Dissolving the support requires the use of chemicals such as sodium hydroxide (caustic soda) that are

Table 5.1 Approximate Properties of FDM/FFF Filaments

Polymer	HDT (°C)	IZOD Impact (J/m)	Tensile Strength (MPa)	Flexural Strength (MPa)
PLA	60	18	49	65
ABS	92	110	34	61
Nylon	82	200	48	69
PEI	152	106	74	116
PPSF	189	59	55	110

highly alkali and therefore care must be taken during their use. The sodium hydroxide is mixed with water in a stainless-steel container and then the solution is heated to 60°C. Some caustic tanks also make use of internal pumps to recirculate the solution around the parts and speed up the support removal. Compared to manual removal of supports there is less chance of damaging the part, and no witness marks remain on the part surface. The time taken to remove the support can vary between an hour for small amounts that are easily accessible up to 12 hours for parts with narrow passages or hollow internal structures.

5.3.1.2 Improving Surface Finish

The relatively large layer thicknesses found in FDM are visible to the human eye, and on parts with curvature in more than one plane these are likely to leave the part with a stepped appearance, not dissimilar to contour lines on a map. These can be removed manually with abrasive paper or a file, and indeed for some materials this may be the only option.

With ABS parts, another option is to intentionally attack the surface of the material with acetone or another solvent. This is relatively safe to use, cheap, and readily available. Parts are placed inside a sealed box, inside which is another open container with some of the solvent. Acetone will evaporate quickly at room temperature; therefore, acetone vapor fills the container and comes into contact with the surface of the part. This begins to attack the ABS, giving it an appearance similar to melted plastic; smooth and glossy. Although the aesthetics are greatly improved and less manual labor is involved, care must be taken not to leave parts in the box for too long as fine detail and crisp edges will also be lost as the process is not selective. This treatment may also have a negative effect on mechanical properties of ABS, so should only be used for nonsafety-critical components.

5.3.2 Powder Binding (3D Printing)

Powder binding, more commonly known as 3D printing, was developed by MIT in 1993. It is the process by which a component is made by laying down thin layers of powder and then using a liquid binder to stick together the particles in the areas in which a solid part is required. The powder used in most systems is a mixture of starch and gypsum plaster, using a water-based binder to activate the plaster. Support structures are generally not required in powder binding because the loose powder itself supports any overhanging features or suspended objects. Without the need for separate supports, both build time and material usage are reduced compared with other AM processes. Any design must consider this feature of the build process if any hollow parts are required; at the very least a small hole must be designed into the part to allow any powder to be released, or it will remain enclosed in the finished part, increasing the cost and weight of the component.

At the end of a build process, the user is left with a cubic container full of loose powder, within which the finished parts can be found. Some commercial systems have an in-built vacuum system that can be used to suck up the excess powder that can be reused, minimizing the amount of waste produced. The binder is usually deposited using a commercially available inkjet printhead meaning it is easy to source a replacement – this also means that an additional color inkjet head is a simple addition, allowing the user to print colored inks at the same time as the binder is deposited on the powder and therefore manufacture parts in full color.

The resulting plaster parts typically lack strength and toughness in their raw state, but various postprocessing methods can be used to improve their mechanical properties. Parts made using this process are suited to assessing form and fit, but in the past had little mechanical strength and were therefore not usable to test function in engineering applications. The latest machines such as 3D Systems' Projet 360 can now use different binders that are selected to give better mechanical properties, color replication, or lower environmental impact (3D Systems, 2015). This means powder binding parts can also be used in limited functional applications.

Many materials can theoretically be made to work using this process, as long as they can be made into a fine enough powder and a suitable binder can be developed to adhere to them. Researchers have used this process to make human bone scaffolds by using a biocompatible ceramic powder and special adhesive binder to make "green" parts that are then sintered at high temperature to make a very strong part (Cox et al., 2015).

Typical applications for 3D printing include full color prototypes for fit and form and architectural scale models. The advantages of powder binding are low build cost, rapid build times, and full color capability. The disadvantages include poor mechanical properties and rough surface finish when compared to SLA or SLS.

5.3.2.1 Postprocessing

The mechanical properties and the brightness of any added coloring can be greatly improved by infiltrating the raw powder parts, for example dipping them into a cyanoacrylate adhesive that improves the bond between the individual powder particles and therefore improves the part properties. Other infiltrants include melted wax and epoxy. These are not expensive if bought in bulk quantities, and the only equipment needed is a container large enough to immerse the component. This is a low-cost and effective way of improving the properties of a part made by powder binding.

Some raw powder parts can be successfully heated in an oven or furnace, either to melt them in the case of thermoplastic polymers, or to sinter them in the case of ceramic powders. This greatly improves their strength, toughness, and porosity. This sintering process can also be used to create alloys of multiple materials if a blend of powders is used.

5.3.3 Powder Bed Fusion (DMLS, SLS, SLM, EBM)

This group of processes make use of a high powered laser or electron beam to locally and selectively fully or partially melt a powdered material together layer by layer to create a 3D object. The beam either fully melts the material in each layer, or applies enough heat for the particles to join together and fuse without ever being fully melted (sintering). The full melt in processes such as electron beam melting (EBM) and selective laser melting (SLM) creates opportunities to change the grain structure and crystallinity of materials, but requires more energy to be transmitted to the build material. Sintering processes include selective laser sintering (SLS) and direct metal laser sintering (DMLS) can result in lower energy, faster build times, and the opportunity to engineer a more porous structure due to the fusion of particles.

A powder bed gradually fills up with the build material as the desired area of powder is solidified through melting or sintering by the energy beam, which is usually directed by mirrors across the desired cross section of material, and at the end of the build the parts have to be removed from this powder. Unlike powder binding, many part geometries will require supports to be generated to prevent part failure, and these will have to be removed from the part postprocess.

Due to the ability of the beam melting process to use intense heat to melt the powders, both polymers (nylon, polystyrene, etc.) and metals (titanium, aluminum, steels including Inconel) can be processed using these systems. Specific systems have even been developed to manufacture parts directly in precious metals such as yellow and rose gold (Cookson Precious Metals, 2014). The key requirement is that the material can be made into a fine powder made up of very round particles of a regular size. This is typically achieved by use of a ball mill and then a range of filters to achieve the required particle size. There is therefore great potential to develop methods for processing new materials and the range of more "official" materials available from the equipment manufacturers is likely to increase.

The latest powder bed fusion systems can achieve material densities of greater than 99%, meaning that the resulting material properties exceed those of equivalent parts made by casting and are approaching those of wrought material. This means that functional, safety critical parts can be manufactured using this group of processes. Metal parts manufactured using this group of processes have the ultimate mechanical properties of any AM component. Engine valves for motorsport have been made in Inconel using DMLS that have the same level of durability as traditionally made components but are 20% lighter (Cooper et al., 2015).

The accuracy and power of the energy beam means that high-resolution parts can be manufactured rapidly, and production times are further reduced by the ability to fill the entire build volume with parts (e.g., build parts one on top of the other).

Due to the laser or electron beam used in this manufacturing process the price of purchasing and maintaining these systems is significant, with prices starting at around £200,000 for the equipment. It is sensible to budget roughly 10% of the cost of the machine for ongoing maintenance in addition to this.

5.3.3.1 Postprocessing

The temperature differential that occurs between completed layers and the material under the focus of the beam results in high residual thermal stresses within the finished component. In extreme cases, large metal parts may even catastrophically fail during the build if support structures are not added in the correct place to prevent excessive thermal stress buildup. Without relieving these stresses the finished component will not achieve its maximum mechanical properties and therefore annealing of parts is recommended.

The support structures for powder bed fusion parts will need to be removed. With plastic parts this can be done by hand using simple tools, but with metal components this may require the use of CNC or electrical discharge machining (EDM).

The surface finish of powder bed fusion parts is rough, and as discussed above there may be witness marks on the component where supports have been removed. To improve the finish, shot peening with glass beads can be used that disguises any machining marks and reduce the surface roughness to as little as 3 μm R_a (Manfredi et al., 2013). If the part needs to be shiny, then normal metal polishing techniques can be used after shot peening.

5.3.3.2 *Other Issues*

Fine metal or plastic powders are potentially explosive and some metal powders such as aluminum are flammable, especially if they are exposed to moisture. There are, therefore, specific health and safety requirements for this type of AM system. To mitigate the risk of fire or explosion, the majority of these machines will operate under a nitrogen or argon atmosphere depending on the material being processed, and specialist powder handling equipment is included with the machine to protect the user when removing parts from the build area. Very high temperatures will be generated in the part during the build process and potentially during postprocessing, so great care should be taken when removing parts from the machine even some time after the process has finished, as metal parts in particular can retain heat for an extended period.

5.3.4 Photocuring (SLA)

Stereolithography was one of the first patented AM processes, and one of the most widely used in the early days of rapid prototyping. The process uses a laser (usually UV) or other light source to cure a thermosetting polymer resin. The resin is stored in a large vat in its liquid state, and the build platform is submerged just underneath the surface. The laser traces each slice of the part onto the platform, curing one layer thickness of resin. The platform then moves down by one layer height and the process repeats. The use of a laser makes this a fast build process, and the accuracy of finished parts is generally very good; their mechanical properties are also moderate to good, and it is possible to produce tooling blanks and other useable parts with this process.

One of the largest build areas found in AM for industrial parts is on an SLA machine; the Mammoth by Materialise can manufacture parts over 2 meters in length and was originally developed for manufacturing large prototype parts such as bumpers for the automotive industry (Materialise, 2015). The cost to manufacture parts of this size is significant and only suited to those with deep pockets!

Unlike powder binding systems, supports are required for SLA parts that are hollow or have overhanging geometry. These are typically generated automatically by software, but will require manual removal after the build is complete that can add labor costs to a part. The photocuring resins used in SLA are relatively expensive and must be stored carefully to prevent premature curing when exposed to light. This, coupled with the large volume of liquid required to carry out a build, is one of the disadvantages of SLA: the tank full of resin represents a significant financial investment, and one which has a limited life cycle and is prone to degradation when subjected to contamination or external light sources. This is less of a problem if large volumes of parts are to be made as the resin will be used up before these issues can occur, and profit will be generated from the finished parts; however, for those wishing to invest in a machine for more occasional use these factors should be taken into account.

Some of these issues have been resolved in recent years by the development of lower cost, small footprint SLA machines such as the Formlabs Form1+ (Formlabs, 2015). These have a modest build volume and therefore use smaller amounts of resin, reducing the amount of material that can go to waste and the investment required. Their initial purchase cost is now comparable to an intermediate FDM system, with prices starting at around £2500.

5.3.4.1 Postprocessing

After the build process is complete, photocured parts will normally be coated in a layer of uncured thermosetting resin. This, along with the support structures, must be removed. This typically involves a process of cleaning parts in a dilute solvent such as isopropyl alcohol (IPA) to remove any excess resin, followed by manual support removal using cutting and abrasive tools.

With some resins, exposure to the UV laser will be sufficient to harden and partially cure the material, but maximum mechanical properties will not be achieved due to the rapidity of the build process and therefore the limited time it is subjected to the laser. For these materials, additional UV exposure can greatly improve the mechanical properties of a component. This can be achieved using specialist equipment made by the material suppliers, and for safety-critical or industry parts this is recommended. However, for those with a smaller budget or making parts that are for form and fit rather than function, some hobbyists have seen good results by using a UV tool sterilizer as used by the beauty industry, or indeed with homemade solutions based on a sealed box and a strip of UV emitting LEDs. Both of these are low in cost and achieve good results.

The key element of the high and low-cost solutions is the even, continued exposure of the part to UV light across its surface to ensure complete curing of the polymer resin.

5.3.5 Material Jetting (MJM, PJP)

Material jetting uses inkjet technology to deposit liquid materials onto a build platform, and then cures each layer using an energy source such as UV light. Each layer can be built faster than with processes such as FDM due to the fact that the machine uses a head with multiple nozzles rather than one. Instead of tracing the outline of each part cross section and then infilling the areas that will be solid, the printhead makes a series of horizontal passes across the whole build area, with material being jetted where it is required. This is similar in concept to using a paint roller instead of a coloring pencil, with material being deposited over a width rather than a single point. Another advantage of using inkjet technology is that multiple heads can be used simultaneously to deposit a range of materials with different properties and even blend them together in much the same way as an office printer blends cyan and yellow inks to make different shades of green. Most MJM machines therefore have more than one printhead to allow them to deposit one or more model materials and a support material, allowing parts with overhangs and hollow structures to be built.

As the resins being deposited are in liquid form, much thinner layers can be achieved than with processes using solid powders or melted filaments. 30 μm is typical with 16 μm possible, giving this process a *Z*-axis resolution up to an order of magnitude better than some FDM processes. Each layer is cured instantaneously by a UV source which is attached to the same assembly as the printheads, and therefore passes over the same area of material that has just been jetted.

The most widely known material jetting process is multijet modeling (MJM) that is a trademark of Stratasys, and is used in the objet range of AM machines.

Material jetting uses a similar method of solidification to photocuring, and therefore faces some of the same challenges in terms of materials. The process is limited to using thermosetting, UV curable polymers that also have the additional limitation of needing to be low enough viscosity that they can be jetted through a tiny piezoelectric nozzle from an inkjet head. While this limits the range of materials available, the properties of those materials can be finely tuned with some clever chemistry so that variation in color, flexibility, thermal stability, and strength are all possible. Coupled with the ability to blend these materials together in fixed ratios during the build process, material jetting is a versatile process. One of the most attractive capabilities from an industrial perspective is the ability to replicate "two-shot" injection molded parts, including those with TPE-like materials. An example of this would be a toothbrush with a rigid plastic body and a rubberized grip. With material jetting this assembly could be manufactured as one component with the different materials effectively bonded together to make one component.

The key limitations for commercially available materials are temperature resistance and strength; the resins are not capable of withstanding sustained high temperatures in the same way that an FDM part made in PPSF would, and their mechanical properties are not comparable with powder bed fusion nylon.

5.3.5.1 *Postprocessing*

Support materials are typically similar to build materials, but formulated to be much softer and more brittle, allowing them to be removed from a component manually or using high pressure water. This need for physical removal requires access to the support material, so care must be taken when designing parts with overhangs or hollow structures to ensure that there is a way to remove the support postprocess.

5.3.6 Lamination (LOM)

Laminated object manufacturing (LOM) is one of the older AM processes, invented in the 1990s, and describes a process that involves the laying down of successive sheets of material that are cut to shape by a blade or laser and then bonded together using an adhesive.

In LOM, the first sheet is applied to the build plate using a roller. The cutting tool (blade or laser) then traces the outline of the part cross section onto the sheet, and then hatches the area of sheet that is not required. The build plate then moves down one layer thickness, and the next sheet is then applied on top of the first. The process is repeated until the part is completed.

The material feedstock is either precut into sheets of a defined size or supplied on a roll and can be paper, plastic, or metal foil, and with or without adhesive preapplied. Normal printer paper is one of the most popular materials as it has a tight tolerance in terms of thickness (which gives the layer height of the finished part) and is low in cost compared to other AM consumables. Laminated parts made using paper have final properties similar to wood and can be sanded, cut, and drilled using normal woodworking tools. An example of this type of system is the Iris by MCor (MCor Technologies, 2014), which uses a sharp metal cutting blade to define the outline of a part, as well as allowing color to be added to the outside edges using a standard color printhead. The use of simple office consumables means that LOM is one of the cheapest AM systems in terms of running costs, and is rapidly overtaking powder binding as the system of choice if full color models for form and fit are required.

5.3.6.1 *Postprocessing*

One of the disadvantages of LOM is the difficulty in removing the supports. Although the cutting tool is normally used to hatch or otherwise weaken the area of material not being used in the build, there is still a significant amount of labor required in breaking off this unwanted material, which sometimes requires use of sharp cutting tools. Although the removal process can be messy, the waste material is normally paper, so this is easily cleaned up and not hazardous.

5.3.7 Process Summary

It is often challenging when considering which AM process to use for a particular application due to the range of materials and equipment types available. The table below provides a quick reference for users to compare and contrast some of the more common AM manufacturing techniques.

5.4 Applications

From the early days of rapid prototyping (RP) to just-in-time direct digital manufacturing (DDM), AM has developed from a niche activity within the wealthiest companies and sectors to a group of technologies that is becoming more accessible to smaller businesses and private individuals. The main uses of AM can be broken down into three key areas: making one-off prototypes, manufacturing end-use parts, and producing tooling that allows for the manufacture of low to medium volumes of components using other manufacturing methods.

5.4.1 Prototypes

The first commercial use of AM was in producing prototype parts to check fit and form, allowing parts to be evaluated before any investment was made in costly tooling or a full production line. The chief advantage of AM for this application is the relative speed with which a part can be produced; there is no need for tooling and only the material required to build the part is deposited, layer by layer. For complex metal components or plastic injection moldings this represented a significant reduction in time and cost for the production of an initial prototype. AM is still widely used for prototyping, and most large manufacturing companies in the automotive and aerospace industries have significant capability to make their own parts in-house.

The capabilities and range of materials available for AM have greatly increased since the early days of RP, with the result that users now have a huge variety of aesthetic and physical properties to choose from when building a prototype part.

5.4.2 Direct Manufacturing

Modern AM systems have increased in accuracy, build speed, repeatability, and material properties with the result that they can be used to produce end-use parts rather than just prototypes. This application is most suited to very complex or high value parts that are to be produced in low volumes (<500). The greatest gains in terms of assembly time and costs are achieved where an assembly of several components is replaced with a single AM part, reducing labor, the number of fasteners, and assembly processes.

An example of AM being adopted by an industry that requires high-performance, safety-critical components is aerospace. Designing and building passenger aircraft is a particularly challenging task. They are high-value products but the production volumes are very low and a continued push for weight reduction and passenger comfort has resulted in a need for intelligent packaging to make the most of the space available. Some of the most difficult items to package are the plastic ducts, which route air conditioning throughout the cabin for passenger comfort. These typically have to fit into tight spaces and take tortuous routes through the cabin space making their geometry very irregular (Figure 5.2). In

Figure 5.2 The complex air ducts found inside a passenger aircraft.

addition, as with all aircraft components, they have to meet the relevant fire safety regulations. They are therefore expensive and difficult to manufacture using most traditional methods, and Boeing took the decision with the 787 Dreamliner to use AM (SLS Nylon) for these components. As well as helping with packaging, ducts can be built in one piece rather than sections, saving on labor costs and assembly time (Davidson, 2012). The aircraft manufacturer had made up to 22,000 3D printed parts by the end of 2013, with 30 going into each Dreamliner; a record for a commercial aircraft (Phillips, 2014).

Its ability to make highly complex parts makes AM ideal for this kind of application, and in this case it would certainly be more cost effective than more traditional manufacturing processes such as injection molding once tooling was taken into account. The use of AM to manufacture end-use parts for aerospace is not limited to polymer processes; metals AM technology is also being adopted by General Electric, allowing them to combine multiple engine components into one piece, saving weight, space, and assembly time (Catts, 2013).

Local Motors, in collaboration with Cincinatti Incorporated and Oak Ridge National Laboratory, used their own process called Big Area Additive Manufacturing (BAAM) to manufacture an integrated chassis and body panels for their Strati electric car (Local Motors, 2015). BAAM uses polymer pellets rather than the filament normally associated with FDM, allowing Local Motors to make use of a wider range of injection molding-grade materials. Moving away from filament also reduces the cost of the polymer feedstock, and by working with polymer supplier SABIC, Local Motors were able to use a carbon fiber-filled ABS to manufacture the Strati's chassis, giving it higher strength and stiffness than a standard polymer AM part (Figure 5.3).

Figure 5.3 The Strati AM car. (Images courtesy of Local Motors Inc.)

Internal features were also able to be incorporated into the body, reducing complexity and reducing assembly time. The vehicle uses the powertrain and suspension from the Renault Twizy city car in combination with the monocoque body to create a functional electric car that was built at the Geneva Motor Show in 2014 to demonstrate the speed of production possible with the BAAM process. Local Motors have set a target of 24 hours to build each monocoque by the end of 2015, down from the 44 hours currently needed.

5.4.2.1 *Medical Parts*

Some industries have embraced AM very rapidly and now make use of the technology to manufacture end-use parts in significant volumes. It surprises many people to learn that the most common component that has been made using AM is the hearing aid; there are more than 10,000,000 in circulation today (Peels, 2014). The ability to customize every hearing aid to perfectly fit into a patient's ear makes them more comfortable and effective, and with AM technology this is no more difficult than manufacturing every hearing aid to be the same.

This has been made possible by the advancement in AM materials, processes, and complementary technologies and software such as 3D laser scanning. An audiologist creates a digital model of a patient's ear canal using a laser scanner. This model is checked and converted into a suitable CAD file complete with interior spaces and fixtures for the necessary electronics before sending it for AM. The relevant components are then assembled into the AM casing before it is fitted to the patient. High-resolution processes are required to ensure a comfortable fit, and specialist materials that have been certified for use in contact with the human body must be used. Compared to the traditional processes used, which involved making wax patterns or using heat to manipulate the shape of a mass produced "average" hearing aid, the AM process is considerably faster and requires less time from a skilled audiologist, which reduces the overall cost. The finished product can be completed in a day and is also a better fit and more comfortable.

The dental industry has also embraced AM, and material development has allowed drilling guides, temporary dentures, and even partial dentures for extended use to be manufactured direct by AM. Companies such as Envisiontec have developed materials and machines specifically for application in the dental industry (Envisiontec, 2015).

5.4.3 Rapid Tooling

One of the key challenges with manufacturing a new product for high volume production is the need to make tooling, which requires significant resource in terms of both time and money. The lead time on tooling for an injection molded plastic part, for example, can be up to 12 weeks and cost many tens or even hundreds of thousands of pounds depending on the complexity and size of the part. Once the tooling is made, however, the cost of each part is typically very low compared to the 3D printed equivalent. Rapid tooling (as the name suggests) uses the relative high speed of manufacture of AM to allow tooling for additional processes to be manufactured much more quickly than traditional methods; the build time for a small insert tool for injection molding can be as little as 4 hours. The key benefit of this process is that it allows a significant number of production parts to be made from a single AM tool. The parts can be made in production intent materials and processes and are therefore truly indicative of what the final parts will be like. The rapid tooling is unlikely to be as resilient or as highly finished as standard tooling, and is likely to survive long enough to produce a few hundred parts or so depending on the material being molded and the geometry of the part.

As well as injection mold and metal press tooling, AM can be used to produce sand sintered molds for casting and patterns for investment casting. The latter are used regularly in the jewellery industry for bespoke items and highly technical pieces that would be very time consuming to produce patterns for using traditional methods. The growth of the use of CAD software for jewellery design has resulted in the ability to produce more complex and intricate designs that require AM to produce the casting patterns. Very high-resolution parts are typically manufactured using machines such as the Formlabs Form1+ or Envisiontec Perfactory range that allow very high-resolution parts to be manufactured in waxes or resins specifically designed to be castable; the tradeoff is a slower build time and a smaller build envelope than that found on more engineering-focussed machines.

As an example, some tooling manufactured at WMG for a small plastic clip took 4 hours to manufacture, cost £400 and lasted long enough to produce 200 components. A quote was obtained from a tool room in the UK for comparison; the point at which it would have been more cost effective to manufacture a steel tool was calculated to be 2400 parts for molding PP and 900 parts for 20% glass filled nylon. This only reflects the cost of the tooling however, and does not quantify the enormous benefit gained by the client from being able to have production ready parts in just a few days after the tooling design was complete. The other major benefit to this process is the ability to make changes to the mold tool design once parts have been made with minimal down time and minimal cost. Making changes to a steel production tool can be very time consuming and expensive – making use of disposable tooling inserts allows rapid development of the tooling to maximize part quality and productivity before the tooling design is approved (Table 5.2).

A press release from Stratasys Inc. in January 2015 announced that consumer goods company Unilever has adopted AM technology to manufacture injection mold tools for prototype parts (Stratasys, 2015). The parts produced are used for consumer testing and checking fit and function in volumes of around 50, with the benefit that they can be made in the final production material (e.g., polypropylene). Unilever claim this has reduced lead times for these preproduction parts by 40%, and tooling changes can be implemented much more quickly than with metal tooling. An Objet Connex MJM machine has been used to produce these tools in a material called Digital ABS for manufacturing a variety of prototype parts such as bottle caps and toilet rim blocks (Figure 5.4).

Table 5.2 Additive Manufacturing Process Comparison

Process	Materials	Resolution	Colors	Surface Finish	Multiple Materials
Extrusion	Most thermoplastic filaments	Moderate/poor	Multiple colors (limited range)	Moderate	Yes
Powder Binding	Limited ceramic and polymer powders	Moderate/poor	Full color	Poor	No
Powder Bed Fusion	Most thermoplastic and metallic powders	Moderate	Single color (limited range)	Poor	No
Photocuring	Limited thermosetting resins	Good	Single color (limited range)	Good	No
Lamination	Paper, plastic sheets	Moderate	Full color	Moderate	No
Material Jetting	Limited thermosetting resins	Excellent	Multiple colors (limited range)	Excellent	Yes

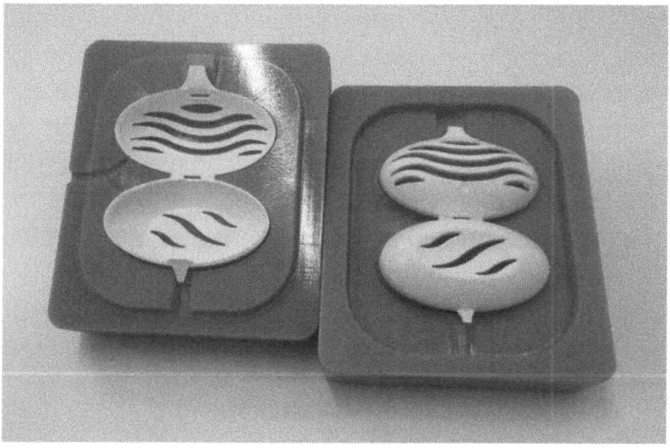

Figure 5.4 An injection mold tool manufactured on an Objet Connex MJM machine (Stratasys, 2015).

An important point to consider if making use of polymer tooling for injection molding or other plastics processing techniques is material compatibility; if you try to injection mold an ABS part using an ABS mold tool there are predictable and unfortunate results! Care must be taken to identify the desired production material and select a tooling material to suit this.

5.5 Multifunctionality in Additive Manufacturing

One of the most exciting recent developments in AM is the ability to vary fundamental component properties throughout the part during the build process. This opens up a whole new range of possibilities for the manufacture of parts with functionalities in addition to their physical attributes. Examples of properties that can be manipulated throughout a part include stiffness, magnetism, electrical conductivity, and the level of alloying elements or fillers.

Multifunctionality is possible with even the lowest cost AM machines by using specialist materials many of which are already available to the market; however, the truly exciting multifunctional printers require novel or additional manufacturing processes. These processes are almost exclusively in the research stage at the time of writing, but given the level of investment in R&D in this field and the current rapid rate of development in AM, these should become market-ready within the next 2–3 years.

Another very exciting area of multifunctionality is called "4D Printing," where the fourth dimension represents time. Using this concept, parts could be made to change their functionality, mechanical properties, or even their physical shape over time under the influence of external sources of energy such as heat, light, or electrical current. In this section, I have tried to describe a range of the most interesting and innovative developments in this area, but make no apology for not covering them all; this is a huge area of research and projects are being proposed every day in universities, companies, and research organizations around the world.

This section has been broken down into commercially available, market-ready technologies and those that are at an earlier stage of development: the future of multifunctional additive manufacturing.

5.5.1 Market-Ready Multifunctionality

One of the simplest ways of increasing the functionality of a part is to add some particles or fibers with different properties to the polymer used as a material feedstock. The growth of open source FDM printers that make use of low-cost plastic filament has allowed greater experimentation to take place at relatively low cost, with the result that many hobbyists and researchers have come up with ingenious ways of functionalizing their builds. Using FDM technology with multiple nozzles, two or more filaments can be deposited during one build process. It is relatively simple, therefore, to produce filaments with differing properties by extrusion compounding and integrate these into one component during the build process. Fiber and particulate additives can, therefore, be integrated into parts at the low cost end of the AM market; producing multifunctional parts is arguably simpler with lower cost machines due to the open source nature of the materials, software, and hardware used.

In higher-end systems the same effects are often achieved by varying the chemical and molecular makeup of the polymer resins used; this particularly applies to systems such as MJM and SLA that use thermosetting liquid resins.

5.5.1.1 Stiffness/Flexibility

"Living hinge" and rubber-like parts have been possible for some time using an Objet Connex printer that makes use of varying formulations of PMMA to make parts with different Shore hardnesses. However, the restriction of using only PMMA has limited the temperature and mechanical performance of the resulting parts. The recent development of polypropylene powders for SLS and flexible TPE filaments for FDM has greatly increased the range of functionalities possible with AM parts. By using PP and TPE, flexible, impact, and fatigue resistant parts can be made while maintaining the superior material properties of those plastics.

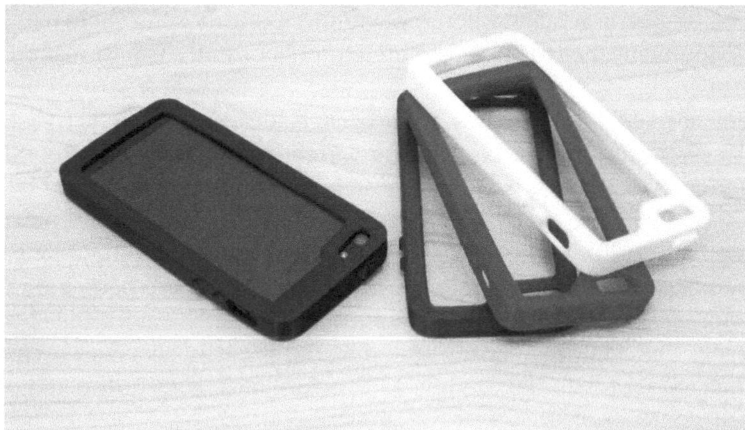

Figure 5.5 A protective mobile phone case made from "Ninjaflex" TPE filament. (Image courtesy of Ninjaflex.)

Ninjaflex is a thermoplastic elastomer (TPE) filament with very high flexibility, allowing owners of FDM machines to manufacture very flexible parts that were previously only achievable using much more expensive equipment (Ninjaflex, 2015). This filament has already been used to make prosthetics and protective cases for electronic devices (Figure 5.5).

5.5.1.2 Electrical Conductivity

Several research groups have identified the potential to selectively create conductive tracks or areas within an additive manufactured part during the build process. This can be achieved in many different ways, either by adding a varying amount of conductive filler to a polymer or by using multiple materials; one dielectric, such as ABS and one conductive, such as silver.

Researchers in the UK successfully compounded carbon black into a polymer material to create a low-cost, conductive plastic called "carbomorph" that is suitable for AM (Leigh et al., 2012). It was possible to achieve loadings of 15% by weight carbon black in the filament while using the standard nozzle on a "Bits from Bytes" BFB3000 low-cost FDM machine. Loadings above 15% resulted in issues with material deposition. Carbomorph was used to make both the plastic body and conductive circuitry for a games console controller, as well as capacitative and flexural sensors all of which were demonstrably functional. By using low-cost materials and equipment, this material has the potential to greatly increase the innovative ways in which AM can be used to manufacture multifunctional parts. Although the researchers did not commercialize this material, the addition of a carbon black filler to a polymer is easily achievable using the same equipment that filament manufacturers currently use to make FDM filaments.

Similar material properties have been achieved with Selective Laser Sintering powders by adding varying weight percentages of carbon (as a powder or short fibers) to nylon. These materials possess greater strength and especially more resistance to temperature than pure nylon powders, making the resulting parts suitable for a wider range of applications. The material was specifically developed for motorsport where parts have to be lightweight, strong, and be able to withstand the harsh environment of the racetrack. The additional cost of the material is easily accepted by motorsport, where part volumes are low and performance is the ultimate goal.

5.5.1.3 Thermochromic Effects (Color Changing with Temperature)

A range of thermochromic ABS and PLA filaments that change color with temperature are now available for use in low-cost FDM machines. There is a growing list of suppliers for these materials (e.g., MakerGeeks, Torwell) with a corresponding increase in the range of available colors and critical temperatures (the temperature at which color change takes place). Although typically used for aesthetic reasons in the hobbyist market, these materials have the potential to be used in many useful applications across sectors such as electronics, clothing, and medical.

Effective cooling and temperature management of electronics can be challenging, particularly in environments that can get dusty and block up any cooling vents in product casings. A case made from thermochromic polymer via AM would not only allow more intricate cooling passages to be created, but also provide a visual indicator of loss of cooling in the event that vents became blocked or any fans stopped functioning.

Wearable sensors have become more widely used in recent years particularly in clothing for elite sportsmen and women. A visual indicator of core body temperature on a garment that was triggered at a precise temperature value could provide valuable information to sports scientists and also be used in the medical industry for patient monitoring.

5.5.2 Tomorrow's Multifunctional AM Technologies

5.5.2.1 Self-Assembly

The term 4D printing has been attributed to Skylar Tibbits, a researcher at Massachusetts Institute of Technology and director of the Self Assembly Lab. In a talk to the TED2013 conference, Tibbits described his vision for parts and materials capable of self-assembly (Tibbits, 2013); the ability to change their physical shape over time when subjected to external influences such as movement, heat, or moisture. MIT have produced demonstrator parts using 4D printing that go from being a straight cylinder to an assembled cube or MIT logo when immersed in warm water. Shape memory polymers have existed for some time, so the idea of a plastic component capable of self-assembly is not entirely new. However, 4D printing allows incredibly complex shapes and assemblies to be made from polymeric materials quickly and simply and with no tooling. Compared to injection molding, the cost of using 4D printing to produce innovative, self-assembling parts is significantly less and the time to market is significantly reduced.

The potential applications for parts with the ability to self-assemble are many, in sectors as diverse as aerospace, medical, and infrastructure. In outer space applications, 4D printed supports could be used to make the exoskeleton of solar panel arrays, allowing them to be packed into a small space for launch, but then to unfurl into very large structures when experiencing zero gravity. This would give a great cost saving due to the smaller, lighter payload needing to be sent into orbit, and also greatly improve the safety of astronauts who would otherwise have to carry out a spacewalk to assemble these structures.

Implantable medical devices such as stents are often difficult to insert and cause a great deal of pain and discomfort for the patient. A self-assembling structure could be made to be very small in diameter for insertion, but then to expand when exposed to human body temperature, thus making the surgery less invasive and improving patient comfort.

The same concept could be expanded to larger pipes used in infrastructure projects to transport water or gas. Pipes often need to be dug up and replaced when the capacity required exceeds the flow rates they can achieve due to their diameter. This causes disruption and carries a large cost. A pipe that could change its diameter on demand could be very efficient when low flow rates were required, but allow for expansion if demand increased.

5.5.2.2 Electrostatic Charge

Researchers have proven the possibility of applying an electrostatic charge to a polymeric material during additive manufacturing that remains in the finished component for an extended period postprocess. This was discovered during a project to use laser printing as an additive manufacturing technique for powder deposition, and has therefore been nicknamed "3D photocopying" (Jones, 2013). The buildup of charge was shown to be detrimental to the quality of the 3D objects being manufactured, but several methods were developed to combat this problem.

Embedding a charge in an object has many potential applications such as adding barcodes or identification marks on a component that are easily detectable using simple equipment but completely invisible to the naked eye. This is attractive to many industries where counterfeiting is a problem, or where lifing of a part is critical for safety reasons. The benefits over traditional barcodes or RFID tags that can be laser etched onto parts or applied as a label include durability as the identification marks are through-thickness and therefore cannot be abraded away and immunity from contamination, as the sensors used are not optically driven. This makes the process suited to parts that may be used in a dirty environment such as the aerospace, automotive, and military sectors.

5.6 Impact of Multifunctionality in Additive Manufacturing

5.6.1 Economic Opportunity

As discussed earlier in this chapter, the AM industry is predicted to be worth $100 billion globally by 2020 (Wohlers, 2012). Much of this growth will be due to the increase in sales of machines and consumables at the lower cost end of the market, which has greatly increased the size of the market for AM by making the technology more accessible to a wider audience. AM is not suited to true mass production so there is little risk to existing high volume manufacturing supply chains.

Multifunctionality is likely to see increased uptake by early adopters in high-value, low-volume applications in sectors such as aerospace, medical, motorsport, and defence.

5.6.1.1 Logistics and Supply Chain

AM makes use of electronic data to define the geometry of finished parts, and the relevant CAD file can be processed by any AM machine's software. The small footprint and self-contained nature of many processes mean that manufacturing using this technology can be truly local, with design work being carried out globally and parts manufactured very close to the area in which they are required. This minimizes transport and logistics costs, and also has the potential to reduce the need for stocks of spare parts to be kept. As long as the digital data is stored electronically, any location within a company can manufacture the required parts by AM. This could be hugely beneficial to the automotive industry that often has to keep tens of thousands of part numbers in stock to cover models that are up to 20 years old, and need to be distributed to a global dealership network.

Multifunctional AM could also benefit from this feature, particularly in applications where it is not possible to replace components that fail using traditional manufacturing methods. An excellent example of an environment that would benefit from this is space exploration; it is not possible to transport traditional manufacturing equipment into orbit due to the weight and the potential danger to astronauts; however, a small, self-contained multifunctional AM machine could enable parts with a range of properties to be manufactured, with any data needed sent to the astronauts from Earth.

5.6.1.2 Time to Market

One of the key advantages of AM is that it requires no tooling, and can therefore manufacture a part much more quickly than traditional manufacturing processes such as injection molding. The ability to produce prototype parts very quickly and therefore prove a design in terms of fit form and function also helps to reduce the time taken to introduce a new product. AM tooling can also increase the speed with which parts can be made using traditional manufacturing processes.

These advantages will translate to multifunctionality in AM, and may have increased impact due to the probable complexity of multifunctional components; to make use of multifunctional materials typically requires more complicated (and therefore time consuming) manufacturing technologies.

5.6.2 Environmental Impact

One of the key benefits of additive rather than subtractive manufacturing is the minimal amount of waste material produced. Rather than beginning with a solid block of material and machining it away to leave behind the desired geometry, AM lays down only the desired material layer by layer. As with all processes there is waste, but it is minimal compared to the total amount of material used. This applies equally to multifunctional as it does to traditional AM, and in fact the reduction in waste may be accentuated by the integration of several functionalities into one component and one build process, which would previously have required several separate manufacturing processes.

5.6.3 Society

Low-cost FDM machines capable of multifunctionality could have a profound impact on areas of the world with little or no access to manufacturing technology. They consume little electricity, are relatively safe to operate, and use cheap, readily available consumables that could be made from recycled plastics. Mobile phone ownership is growing very quickly worldwide, and together with the drop in price of smartphones and availability of CAD/design mobile apps, there is the potential for users in remote areas to be able to manufacture useful components without the need for expensive equipment and with only minimal training.

5.7 Barriers to Multifunctional Additive Manufacturing

In order for multifunctional AM to be adopted, there are several barriers that must be overcome. These relate to all aspects of the technology lifecycle from intellectual property of the initial idea, to the software needed to the treatment of multifunctional parts at end of life. In this section, each of these barriers is described, along with potential routes to mitigate or overcome these barriers.

5.7.1 Environmental Issues

For AM parts that gain their multifunctionality through combination of different materials, consideration of product end of life is essential. As with any component with a mix of materials, effective separation of each individual material is required before they can be recycled. This is a particular challenge for those processes that add an infinitely variable amount of filler to the build material during the manufacturing process as the filler will be very difficult if not impossible to separate out, and could be found in different concentrations in each part. For those processes that can combine many different materials, for example an FDM machine with several nozzles that can process filament with different additive content, material traceability, and therefore the ability to recycle parts is challenging.

In order to mitigate this barrier to adoption, changes can be made to the way in which components are designed. While many plastic parts made for the consumer are disposable or have a relatively short life, multifunctional AM or 4D printing should be used for parts that have a much longer lifetime, or those that integrate several functionalities and therefore remove the need for more than one component, thus reducing the overall environmental impact of the assembly.

For applications where recyclability is a key attribute, processes that make use of a single material should be selected in order to simplify segregation of materials for recycling at end of life. If multiple materials are combined and cannot easily be separated, specialist advice should be sought to enable them to be disposed of safely. This may incur additional costs, either through landfill taxes such as those encountered in the UK and Europe, or through charges made by specialist disposal companies if the material contains a potentially harmful substance.

5.7.2 Intellectual Property

A potential barrier to any new area of technology is the protection of intellectual property. If patents are filed to protect an innovative area, technology development is limited in this area for all but the patent holder. AM is already an industry dominated by patents and design protection, and although this has encouraged researchers to invent new processes and equipment to get round this protection, it still presents a potential barrier to the future development of novel equipment capable of multifunctional AM.

IP protection, however, can also be a method of preventing a technology being monopolized as demonstrated by researchers on the RepRap project. The open hardware license agreement means that anyone can use the basic design to create their own machine to sell commercially, but they cannot patent their design. This has enabled a huge range of machines to be developed and arguably resulted in the greatest level of innovation in AM (albeit at the low cost end of the market) since the late 1980s.

5.7.3 Material Supply and Compatibility

Currently, the most accessible route to innovation with AM materials (outside of the major manufacturers) comes from use of the lower cost, open source machines. Their low price point means that there is little financial risk for researchers if a trial goes catastrophically wrong, and their ability to accept a wide range of materials means that researchers are able to manufacture their own filaments, resins, or powders and use them to construct parts.

Many of the AM machines at the higher end of the market require users to make use of the manufacturers' own materials to produce parts to avoid violating the warranty. The more cynical among us may believe this is to elevate the profit to be made from creating a monopoly on supply of materials, but another point to consider is that the reliability and accuracy of these production systems relies on both the quality of engineering of the machine and the materials it uses. Any manufacturer would quite reasonably wish to avoid a situation where damage had been done to a machine through use of substandard materials or an additive that reacts badly with some of the materials used in its construction. In FDM for example, the diameter and formulation of the filament used has a direct bearing on the density, strength, and appearance of the finished part.

Any development of multifunctional materials should ideally be in partnership with the equipment manufacturer to ensure the quality and reliability of the resulting products are maximized. This requires the buy-in of the major equipment manufacturers and a more open approach to material and process development. An example of how this process might work is HP's "Open Customer Engagement Program," which they have introduced for their Multi Jet Fusion technology, due for commercial release in 2016 (REF). By working with selected partner organizations and providing access to the core technology, they are enabling external groups to develop innovative materials that will increase the capabilities of their technology, opening up the potential applications and functionality that can be built into their AM machines. This model of releasing technology as a "beta" version before it is commercially available to enable collaboration and innovation is widely used in software development. Microsoft released the Kinect motion tracking camera system for their Xbox 360 console along with full access to the software that operated it months ahead of its general release, enabling selected research partners to come up with novel ways of modifying and using the technology.

5.7.4 Software and File Formats

3D design and CAD software were previously the domain of manufacturing companies and were therefore prohibitively expensive for the individual user. The cost of CAD software has reduced over the last decade, and various companies offer low-cost licenses for use in education and research or monthly licenses that reduce the initial outlay for those starting out in CAD. There are also a growing number of free software packages such as Sketchup and 123D that give an opportunity to generate files of some complexity without purchasing software. The barrier to multifunctionality here is the level of skill and knowledge of the manufacturing process that is required to create a successful part at the end of the AM process flow. Industrial CAD packages require a significant level of training and experience to use them to their potential, and are more suited to generating parts for traditional manufacturing processes. More freeform design software allows more freedom for the designer with fewer constraints on how the user creates features in the 3D model. Care must be taken with this type of software to ensure that the model that has been created does not exceed the limitation of the final AM process.

A core element of AM is the basic .STL file format used by most commercial machines, and the way in which this is processed by software packages (see Figure 5.1). The .STL is a polygonal mesh of the surface of a part, and is therefore not a solid model. Editing an .STL file using most commercial CAD packages is, therefore, very difficult (if possible at all), and specialist software is required that comes with a price tag in the tens of thousands of dollars for a license. This inflexibility in file formats presents a potential barrier to multifunctionality. It may be desirable to produce a part made up of many different materials, or a continuously varying amount of an additive throughout its structure. The limitation preventing this part being made is unlikely to be the manufacturing process itself, but rather the production of the 3D CAD file that describes it and the way in which that file is converted into toolpaths and machine code.

Once a CAD model has been produced and converted to a format suitable for AM, the slicing software unique to every AM machine will need to be capable of dealing with any additional process parameters.

5.8 Conclusions

Additive manufacturing presents a significant opportunity for those looking to embed multifunctionality into components. Although a mature set of technologies, recent media interest, and the expiry of patents have led to a huge growth in the rate of development of lower cost systems that are readily accessible and easy to modify. Innovation in materials has also enabled a range of properties and functionalities to be built into parts during the AM process.

Improvements in strength, economics, build speed, and functionality have made AM suitable for the manufacture of end-use components, not just early stage prototyping. The technology has been adopted for applications as diverse as aircraft components and hearing aids, and exciting developments have been made in the use of AM to produce injection mold tools, greatly reducing the time to market for low-volume molded plastic parts.

Although not suited to mass production AM provides a unique opportunity to design unique parts to order, making it an ideal way to make high value, complex components more accessible. The most influential barriers in this field are not unique to AM; finding ways to safely deal with materials at end of life and ensuring a regular and reliable supply of raw materials are a challenge for any process that delivers multifunctionality. Specific to AM, the need for software to be developed hand-in-hand with new materials and processes presents a challenge to the industry, as does the need to prevent intellectual property protection from stifling innovation.

These barriers are arguably outweighed by the significant benefits that could result from more widespread development of this technology. The AM industry is experiencing rapid growth that is predicted to be sustained for the next decade as more materials, machines, and processes are developed, making it a popular target for investors. The economic opportunity presented by adding functionality to components and therefore increasing their usefulness and potential markets is significant and likely to contribute to the growth of the industry more as the market for lower cost AM machines becomes saturated. Other advantages for industry include the shortening of supply chains by providing local AM manufacturing capabilities to manufacture parts on demand, and the great reduction in time between design and manufacture that is made possible by the lack of need for tooling.

It is already possible to create parts that change color with temperature and have different mechanical properties throughout their structure using AM, and in the future we are likely to see components made on demand that can change their shape, be easily tracked using invisible tags, and even put themselves together.

Few areas of technology are receiving as much investment or research focus, and the move toward multifunctional additive manufacturing is likely to create a step change in the way we think about manufacturing complex parts.

List of Acronyms

3DP 3D printing
ABS Acrylonitrile butadiene styrene
AM Additive manufacturing
ASTM American Society for Testing and Materials
BAAM Big area additive manufacturing
CAD Computer aided design
CNC Computer numerical control
DDM Direct digital manufacturing
DMLS Direct metal laser sintering
EBM Electron beam melting
FDM Fused deposition modeling
FFF Fused filament fabrication
LED Light emitting diode
LOM Laminated object manufacturing
MIT Massachusetts Institute of Technology
MJM Multijet modeling
PEI Polyetherimide

PJP Plastic jet printing
PLA Polylactic acid
PMMA Polymethylmethacrylate
PP Polypropylene
PPSF Polyphenylsulfone
R&D Research and development
RepRap Replicating rapid prototyper
RP Rapid prototyping
SLA Stereolithography
SLM Selective laser melting
SLS Selective laser sintering
TPE Thermoplastic elastomer
UV Ultraviolet

References

3D Systems, 2015. ProJet 360 Full Colour 3D Printer [Online] Available from: <http://www.3dsystems.com/3d-printers/professional/projet-360> (accessed February 2015.).

ASTM, 2012. Standard Terminology for Additive Manufacturing Technologies. ASTM F2792-12a. ASTM International, USA.

Catts, T., 2013. GE turns to 3D printers for plane parts [Online] Available from: <http://www.businessweek.com/articles/2013-11-27/general-electric-turns-to-3d-printers-for-plane-parts> (accessed January 2015.).

Cookson Precious Metals, 2014. The first Direct Metal Laser Sintering technology designed for the jewellery industry [Online] Available from: <http://www.cooksongold-emanufacturing.com/> (accessed 25.11.14.).

Cooper, D., Thornby, J., Blundell, N., Henrys, R., Williams, M.A., Gibbons, G., 2015. Design and manufacture of high performance hollow engine valves by additive layer manufacturing. Mater. Des. 69, 44–55.

Cox, S.C., Thornby, J.A., Gibbons, G.J., Williams, M.A., Mallick, K.K., 2015. 3D printing of porous hydroxyapatite scaffolds intended for use in bone tissue engineering applications. Mater. Sci. Eng. C 47, 237–247.

Davidson, P., 2012. 3-D printing could remake U.S. Manufacturing, USA Today [Online] Available from: <http://usatoday30.usatoday.com/money/industries/manufacturing/story/2012-07-10/digital-manufacturing/56135298/1> (accessed 07.07.12.).

Envisiontec, 2015. Our Vision – The Fully Functional Digital Dental Lab. Available online from: www.envisiontec.com.

Jones, J.B., 2013. Investigation of laser printing for 3D printing and additive manufacturing, PhD Thesis, WMG, University of Warwick, UK.

Jones, R., Haufe, P., Sells, E., Iravani, P., Olliver, V., Palmer, C., Bowyer, A., 2011. RepRap – the replicating rapid prototyper. Robotica (29), 177–191.

Leigh, S.J., Bradley, R.J., Purssell, C.P., Billson, D.R., Hutchins, D.A., 2012. A simple, low-cost conductive composite material for 3D printing of electronic sensors. PLoS One 7 (11), e49365.

Local Motors, 2015. Available from: <http://localmotors.com/3d-printed-car/>

Manfredi, D., Calignano, F., Krishnan, M., Canali, R., Ambrosio, E.P., Atzeni, E., 2013. From powders to dense metal parts: characterization of a commercial AlSiMg alloy processed through direct metal laser sintering. Materials (6), 856–869.

Materialise, Mammoth stereolithography [Online]. Available from: <http://manufacturing.materialise.com/mammoth-stereolithography> (accessed February 2015.).

MCor Technologies, MCor Iris – True colour 3D printer [Online]. Available from: <http://mcortechnologies.com/3d-printers/iris/> (accessed 26.11.14.).

NinjaFlex, 2015 [Online]. Available from: <http://www.ninjaflex3d.com/>

Peels, J., This is the most successful 3D printed product in the world, Inside 3DP, Online, Published September 22, 2014, Available from: <http://www.inside3dp.com/successful-3d-printed-product-world/>

Phillips, S., 5 Companies Using 3D Printing to their Advantage, Inside 3DP, Online, Published July 21, 2014, Available from: <http://www.inside3dp.com/5-companies-using-3d-printing-advantage/>

Stratasys, Unilever Leverages 3D Printing Injection Molds, Slashing Lead Times for Prototype Parts by 40%, Press Release, January 20, 2015, Stratasys Ltd., Minnesota, USA.

Tibbits, S., The Emergence of 4D Printing, TED2013 Conference, February 2013, Long Beach, FL, USA, Available online from: http://www.ted.com/talks/skylar_tibbits_the_emergence_of_4d_printing?language=en

Wohlers, T. (Ed.), 2012. Wohlers Report 2012: Additive Manufacturing and 3D Printing State of the Industry. Annual Worldwide Progress Report. Wohlers Associates Inc, USA, ISBN: 0-9754429-8-8.

Wohlers, T. (Ed.), 2014. Wohlers Report 2014: Additive Manufacturing and 3D Printing State of the Industry. Annual Worldwide Progress Report. Wohlers Associates Inc, USA, ISBN: 978-0-9913332-0-2.

6 Future Outlook

Vannessa Goodship

6.1 The Future of Manufacturing Multifunctional Systems

This book has looked at the various ways that functionality can be introduced into a plastic product. This is summarized in Figure 6.1.

The first of these options is to embed functionality. A variety of potential materials for this option can be found in Chapters 2 and 3, and tend to fall under the more general term of multifunctional composites. This kind of technology enables the use of materials that are electroactive, antimicrobial, or thermally conductive for instance. Shape memory polymers also fall under this category; the shape function is already embedded into the materials themselves. This type of functionality is very much about the materials themselves and their formulations. Beyond the materials preparation stage carried out by extrusion (Chapter 1) or deposition methods (Chapter 2), these materials are essentially readymade from a manufacturing perspective.

Addition of functions is where the use of additional or different manufacturing routes can be utilized to introduce further functional layers into a system; for example, through processes such as coextrusion blow molding, in-mold films, and additive manufacturing. Most relevant chapters here are Chapters 3, 4, and 5. In-mold films provide a relatively simple and technologically mature add-on route to greater functionality. Full integration of this stage directly into manufacturing is not too far away and can already be seen in some applications. This can also perhaps be considered a through way for future products to enter the next stage of functionality.

The third option, "integration of functions shared in a volume of material," is where the material operates as a fully integrated structure and/or product. This concept is also referred to as the development of multifunctional structures (MFS). It is this option that most excites scientists and manufacturers alike as the ability to create structural batteries in aircraft wings, transparent bridges, clothes that power our mobile phone, or milk cartons that order a replacement for you when empty, can be fully realized.

These types of structures are essentially entire systems in one component, and have the potential to reduce material usage, manufacturing costs, product mass and still communicate across the internet!. A simpler

Design and Manufacture of Plastic Components for Multifunctionality
http://dx.doi.org/10.1016/B978-0-323-34061-8.00006-5

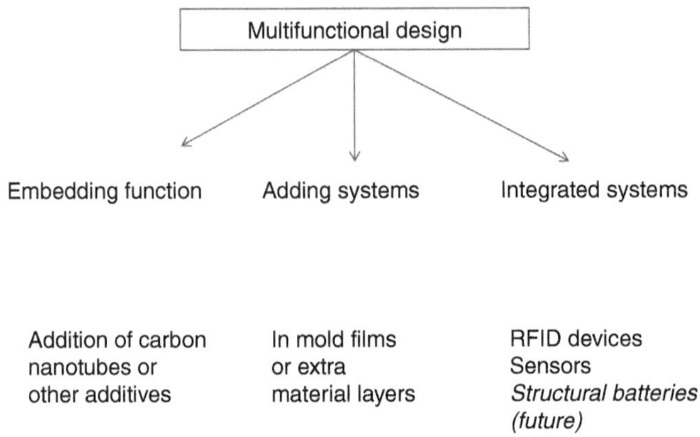

Figure 6.1 Options for multifunctional design.

example was introduced; an embedded antenna that shares the mechanical load such as shown in Chapter 4.

Throughout this book it is shown how material and processes have developed over time with new hybrid innovative processes springing up. For example, inkjet printing is currently being developed as next-generation additive manufacturing where deposited layers could be built of different materials and could be printing products such as low-cost photovoltaic cells and battery structures. These have already been achieved, at noncommercial operating levels, in research institutions.

It should also be clear how changes in the electronics sector and particularly the rise of conductive polymers are now impacting all plastic manufacturing fields whether composite, processing, or additive manufacturing; the drive toward electrically active multifunctionality is striking. However, this is unsurprising given the other technological drivers of automation and the internet of things that is perpetually driving manufacturing and innovation forward.

However, there are other counter arguments against the pull toward such systems, not least of these is the recycling such complex and integrated systems. The development of recycling systems traditionally tends to lag behind waste streams, not least because it takes a while for waste steams to go from product to end of life. The rise of electronic devices and electronic waste has, therefore, driven the need for new solutions and new legislation in the European Union to cope with fast product turnovers and new kinds of electrical and electronic waste streams.

The temptation to go for full MFS must also be offset against considerations in the full life cycle analysis, not just the environmental burden but also any economic and social impacts.

Consideration of designing for recycling at these early stages of product conception will reduce the subsequent burden for all in the future.

However, the bottom line for any manufacturing market has been and always will be unchanged. Can it be made for a profit, if the answer is yes, chances are ways can and will be found to bring such products to market.

This is where partial integration can often be used to allow more cost-effective products to be introduced. This reduces manufacturing stages, realizing better costs and energy efficiency of both processing and products. There is also a reduced complexity in comparison to designing entire fully integrated new systems that may require step-change analysis and rapid learning within the workforce. Evaluation of all potential multifunctional manufacturing routes is, therefore, essential and road mapping the development of future products against current and pipeline manufacturing routes can be undertaken. Such step-wise changes were outlined in technological developments for in-mold labeling in Chapter 4 as products built up in complexity over time.

Disruptive manufacturing routes can appear at any time; the current rise of additive manufacturing is one such example of a production route, which is now free from patents that restricted its developmental potential, but will it mean the end to traditional plastic and composite manufacturing systems?

While additive manufacturing does have many advantages over traditional production processes in terms of bespoke manufacture, as a mass manufacturing system to compete with injection molding or structural composite manufacture, there is still a very long developmental road ahead. This is, however, where the three processes can come together in helping to realize multifunctional designs as each has a role to play in manufacturing the complex products of the future.

Consider the three boxes in Figure 6.1 again, and now consider which processes can best meet the demands of embed, addition, and integration.

In the authors' opinion, there is, and as likely always will be, a broad diversity in plastic manufacturing platforms and it is this that allows plastics to fully realize the incredible variety of properties that are available to today's and tomorrow's designers.

There is more design freedom, customization potential, and cross over manufacturing integration potential for plastics than ever before, and this continues to grow rapidly as manufacturers pursue economic profitability and low costs.

This is now the age of multifunctional plastic design.

Index

T

FSC
www.fsc.org
MIX
Papier | Fördert gute Waldnutzung
FSC® C083411

Kolibri 360 GmbH
Ferdinand-Jühlke-Straße 7
99095 Erfurt, Deutschland
produktsicherheit@kolibri360.de